T0221558

Worlds of Gray and Green

CRITICAL ENVIRONMENTS: NATURE, SCIENCE, AND POLITICS

Edited by Julie Guthman and Rebecca Lave

The Critical Environments series publishes books that explore the political forms of life and the ecologies that emerge from histories of capitalism, militarism, racism, colonialism, and more.

Worlds of Gray and Green

MINERAL EXTRACTION
AS ECOLOGICAL PRACTICE

Sebastián Ureta and Patricio Flores

UNIVERSITY OF CALIFORNIA PRESS

The publisher and the University of California Press
Foundation gratefully acknowledge the generous support
of the Ralph and Shirley Shapiro Endowment Fund
in Environmental Studies.

University of California Press
Oakland, California

© 2022 by Sebastián Ureta & Patricio Flores

Library of Congress Cataloging-in-Publication Data

Names: Ureta, Sebastián, author. | Flores, Patricio, 1985– author.
Title: Worlds of gray and green : mineral extraction as ecological practice /
  Sebastián Ureta & Patricio Flores.
Other titles: Critical environments (Oakland, Calif.) ; 11.
Description: Oakland, California : University of California Press, [2022] |
  Series: Critical environments : nature, science, and politics ; 11 | Includes
  bibliographical references and index.
Identifiers: LCCN 2021057668 (print) | LCCN 2021057669 (ebook) |
  ISBN 9780520386280 (cloth) | ISBN 9780520386297 (paperback) |
  ISBN 9780520386303 (epub)
Subjects: LCSH: El Teniente (Mine)—Environmental aspects. | Copper mines
  and mining—Environmental aspects—Chile. | Water—Pollution—Chile—
  Carén Reservoir Region. | Carén Reservoir Region (Chile)—Environmental
  conditions. | Human ecology—Chile. | BISAC: SCIENCE / Environmental
  Science (see also Chemistry / Environmental) | SOCIAL SCIENCE /
  Human Geography
Classification: LCC TD428.C66 U74 2022 (print) | LCC TD428.C66 (ebook) |
  DDC 363.739/40983—dc23/eng/20220104
LC record available at https://lccn.loc.gov/2021057668
LC ebook record available at https://lccn.loc.gov/2021057669

Manufactured in the United States of America

31 30 29 28 27 26 25 24 23 22
10 9 8 7 6 5 4 3 2 1

CONTENTS

FIGURES

PREFACE: BEYOND EXTRACTIVISM

The results of the election of May 17, 2021, were astonishing for many in Chile. Aimed at selecting members for the assembly to be put in charge of drafting a new constitution for the country, the election was one of the most prominent consequences of the massive social movements started on October 2019. Triggered by an increase in fares for the public transport system of Santiago, the movement—known as Estallido (outburst) or Revuelta (revolt)—included the most massive rallies in the country's history, extensive destruction of public infrastructure, and multiple casualties due to police violence, completely paralyzing the country for several weeks. Motivated by the inequalities and indignities resulting from Chile's more than thirty years of adherence to a radical neoliberal governance program, the movement was only (somewhat) appeased when members of the parliament agreed to hold a plebiscite exploring the possibility of changing the country's constitution. Drafted during Augusto Pinochet's dictatorship, the constitution has traditionally operated as the backbone of *el modelo* (the model), as the neoliberal system is known. After voters overwhelmingly supported the proposal in October 2020, the election of the members of the assembly was the last step before actually starting the process of drafting a new constitution.

Many participants—especially those who had been directly involved in the revolt—were deeply suspicious of this process. Although it was publicly labeled as utterly democratic, the introduction of technical barriers made it likely that representatives from traditional political parties (and the elite groups supporting them) would win most of the seats in the assembly. Vote counting showed, however, that a substantive number of independents and progressives were elected to the assembly, most occupying an elected position for the very first time. As a consequence, and for the very first time in Chile's history, the constitution would be drafted by a truly heterogeneous group of citizens, representing multiple different constituencies and regions, ethnicities, genders, and cultures.

Despite their diversity, most of the newly elected members of the constitutional assembly agreed about the need to fundamentally redefine Chilean society's relations with the natural environment. A keystone for such redefinition was the need to, as it was explicitly stated, "move beyond extractivism," meaning abandoning the relentless extraction of natural resources upon which the Chilean economy had rested for decades.[1] This model has left behind a pervasive legacy of economic inequality, material destruction, and environmental injustice. Moving beyond extractivism implies recognizing the violence inflicted upon the human beings living in the so-called sacrifice zones surrounding extractive projects, as well as establishing a relationship of respect with nonhuman beings and the earth as a whole, something especially critical in a time of climate change. By far the largest extractivist endeavor in the country, the mining industry, could expect to be importantly reshaped as a result of the application of this principle.

Such a transformation presents multiple challenges for a country like Chile. They are already evident in the parallel commitment to continue pursuing the, in the words of representatives of a group of leftist members of the assembly, "economic and territorial development" of the country. Since its early adoption, the concept of development in Chile has been intimately tied to extractive industries, especially mining, and all its negative socio-environmental consequences (as this book explores). If development is still an aim, some degree of resource extraction will continue to be necessary. After all, even the ongoing transition toward sustainable modes of production and consumption rests on the extraction of minerals such as lithium, and Chile is one of its most important producers worldwide. If the new constitution aims at combining deep environmental commitments with some sort of development project, it will be forced to engage in the massive challenge of learning to *practice extraction without extractivism.*

To imagine extraction without extractivism requires us to radically rethink the ways we engage with nonhuman entities, especially minerals. Instead of seeing them as inert and singular materials, we will have to start seeing them as entities entangled in multiple relations, as vital components of several ecologies. Some of these ecologies produce value, mainly through chains of transnational interchange. Others, many others, enact multiple forms of damage, through dispossession, pollution, and violence. Yet others produce neither value nor damage, but perform completely new arrangements of the living. In this book we explore some of these multiple ecologies of extraction.

Our motivation is not only describing these entanglements, but also developing a conceptual apparatus to think about extraction in different ways, beyond the usual for/against models. *Practicing* extraction differently—moving "beyond extractivism"—can only start by *thinking* of extraction differently. We sincerely hope that in the constitutional moment on which Chile is currently embarked, these novel ways to think about extraction can contribute to the urgent task of devising more just and mutually caring ways of living as a nation in our damaged world.

ACKNOWLEDGMENTS

This is not the book we were planning to write, at least not at the beginning of this project in 2012, when a grant was secured to study the governance of mining waste in Chile. As tends to happen, our wide interest progressively narrowed down. First, and after making contacts with several institutions and corporations, CODELCO's El Teniente mine agreed to open its doors to us to carry out fieldwork. From an initial inquiry about the waste management of the mine, we became progressively more interested in what happened at its current tailings dam, Carén. Finally, among the multiple things happening there, the interactions between tailings, water, and a plethora of biological entities increasingly caught our attention. These were the stories that clamored more loudly to be written, the ones that followed us insistently during the multiple days passed at the mine and its environs and while transcribing fieldnotes and collecting documents later on. These stories haunted us, to the point of feeling that we had no option but to write them down in the shape of this book. For this reason, our first acknowledgment is to the multiple beings, from algae to human beings, who lent us these stories. We sincerely hope that—in the case they could somehow read this book—they would not feel that we have let them down, that we have done justice to their travails and worries, their pains and desires.

We would like to thank the institutions that have housed and funded us while writing these stories. The fieldwork on which this book is based started in 2012 when Sebastián had just arrived to work at Departamento de Sociología, Universidad Alberto Hurtado, Chile. Both the university and our colleagues there have been a fundamental support all along the way, especially in terms of giving us time to carry out fieldwork, analyze the material collected, and write this manuscript. We also appreciate the support and comments of several colleagues in the department in the process. Most of the fieldwork was funded by two FONDECYT grants from ANID, Chile's National Agency for Research and Development (grant numbers 1130156 and 1170153).

While still in the initial stages of this process Sebastián was lucky to spend some time as a Carson Fellow at the Rachel Carson Center for Environment and Society in Munich, Germany. The lively RCC community (and its library!) provided a rich environment for the maturation of several key ideas that would guide the writing of this book. Later on he had the opportunity to spend time as a visiting fellow at the Max Planck Institute for the History of Science (MPIWG) in Berlin, Germany, thanks to an invitation from Wilko Graf von Hardenberg. The work carried out there with Wilko and Thomas Lekan, focused on the study of environmental baselines, was also instrumental for some key notions of this book.

Multiple friends and colleagues were involved in different stages of this project. First of all, Sebastián would like to thank Tomás Ariztía, José Ossandón, Ignacio Farias, Manuel Tironi, Matias Bargsted, and Ignacio Arnold for all the years of friendship, intellectual debate, and dark humor. We especially appreciate the help of Manuel, who dedicated time to reading and commenting on the whole manuscript, giving us key insights for improving it. We also appreciate the discussions about the book's contents, and related topics, provided by Abby Kinchy, Arn Keeling, Cristobal Bonelli, Martín Arboleda, Javiera Barandiaran, and Endre Danyi. Elizabeth Povinelli and Soraya Boudia and coauthors were kind enough to share with us the final drafts of their forthcoming books, which were highly influential in drafting the final version of the manuscript. We would like to also recognize Abby Kinchy, Roopali Phadke, and Jessica Smith for organizing the wonderful STS Underground meetings, providing a platform for our acquaintance and interchange with a wide community of people interested in studying the extractive industries from an STS perspective.

We also appreciate the work of Professors Julie Guthman and Rebecca Lave, editors of the Critical Environments book series of University of California Press, who provided enthusiastic support for this project from our very first inquiries. We are especially thankful to Rebecca, who also took the time to provide us with insightful comments on the whole manuscript. We would also like to warmly thank Stacy Eisenstark, book editor at University of California Press, for her help and guidance all along the process of turning our draft into a proper book. We are also indebted to the University of California Press personnel involved in the process: Naja Pulliam Collins, Teresa Iafolla, and Kate Warne. We would also like to express our gratitude to Jon Dertien and Sharon Langworthy from BookComp for their copyediting efforts on the manuscript. We also acknowledge the illuminating comments on the draft provided by two anonymous reviewers.

The fieldwork on which this book is based would not have been possible without the help of several research associates and assistants who participated in various stages from its start in 2013. We are especially indebted to Consuelo Biskupovic, who occupied a central role in the project early on. We would like to thank Francisco Godoy and Iván Sandoval, who were research assistants at an early stage of the project. We also want to thank María Hinojosa and Gabriela Flores for making some of the maps and illustrations for us, and Alejandrina Pinto, who was willing to share with us some pictures from her blog Recuerdos de Loncha.

We would like to give thanks centrally to all the people who participated in this study. First, we would like to offer our deepest thanks to the people at Division El Teniente of CODELCO Chile, who agreed to let us carry out fieldwork on the mine's premises, besides facilitating interviews and obtaining documents. In particular we would like to thank both the heads and the *viejitos* of the Tailings and Water Management unit at Teniente for all their goodwill in allowing us to accompany and (usually) pester them on their daily chores. Second, we would like to thank all the experts, consultants, and government officials who were interviewed during this project. Finally, the openness and good disposition of the inhabitants of the Carén Creek basin was centrally important for the project, allowing us to get a sense of what it is like to live and work in the shadow of a dragon. Without their generosity and openness this book would not have been possible.

A modified section of chapter 2 appeared as a paper in *Environment and Planning D: Society & Space* under the title "Don't Wake Up the Dragon! Monstrous Geontologies in a Mining Waste Dam" (36, no. 6 [2018]: 1063–80). We thank Sage Publications for allowing republication of excerpts from that paper.

Finally, we would like to thank our families for all their support throughout the years. During the process of researching and writing this book, Sebastián has been accompanied by his partner Sylvia. It is certainly an understatement to say that she has been essential for the success of the project. Without her support and love, the countless ways in which she makes his life better, this book would simply not exist. Their children Lucía and Manuel have not only given Sebastián constant joy and support, an alleviation from the usually lonely task of writing. They have also offered the key motivation to ask the difficult questions behind the book, knowing that the damaged worlds we explore are theirs to inhabit, much more than ours.

On his part, Patricio thanks his family for their comprehension and loving support throughout his career as a young researcher. Especially, he thanks Carolina, who was with him patiently while he worked on this project from a small room in Coventry, United Kingdom. He also is indebted to all his friends, particularly Renato and Alejandro, whose pessimism about the current environmental crisis provided the necessary stimulus to take part in the seemingly impossible task of looking for signals of hope in a ruined world. Finally, he expresses his deepest gratitude to his nephew, José, who, like many other children of today, will face the challenge of living in catastrophic times. Hopefully this book can inspire them to find and nurture traces of life in the middle of the ashes.

Introduction

No hay otro tranque de relave en Chile con las características de
Carén, que, por lo demás, es de CODELCO. (There is no other
tailings dam in Chile with the characteristics of Carén, which,
besides, belongs to CODELCO.)

ANA LYA URIARTE, *CONAMA director, July 7, 2006*

GRAY WORLDS

It took us awhile to reach the top of the wall. The dirt road leading us there
started at ground level, climbing for a dozen meters until reaching a flat
stretch that marked the maximum height of an earlier stage in its construction process. It then climbed another dozen meters to a new flat stretch. It
continued thus until reaching its current summit, from where we could see a
panoramic view of the entire basin. Although only half its projected final size,
the wall is already massive. Made of rocks and dirt, it is more than 90 meters
high and a kilometer wide. We drove along it for a few minutes, finally stopping at its middle section, with the front of our truck facing east toward the
millions of tons of tailings already accumulated there. It was January 2014,
and we were visiting for the first time Embalse Carén, the current tailings
dam of Mina El Teniente.[1]

Property of the public company Corporación Nacional del Cobre
(CODELCO), Teniente is located 100 kilometers south of Santiago, Chile's
capital city, in the Andes Mountains. Sitting on "the world's largest known
porphyry Cu deposit," the mine has been in industrial operation since the
beginning of the twentieth century and currently comprises more than
3,000 kilometers of tunnels; it is usually considered the largest subterranean
mine in the world.[2] The mine produces mostly copper (470,000 metric tons
in 2018), among other refined minerals. As usual in contemporary mining,
each copper ton involves the parallel production of around 99 tons of tailings, residues that need to be removed rapidly from the processing plant in
order to avoid clogging the system. Immediately after being produced in

FIGURE 1. Eastward view from the Carén Dam wall

massive decantation ponds, Teniente tailings are put into a concrete canal, through which they flow for 85 kilometers across Chile's central valley until they reach Carén Dam, located in a valley in a mountain range known as Altos de Cantillana.

The view was truly impressive. In the area near the wall (see figure 1), tailings were covered by a large body of bright green water, stretching for several kilometers into the distance to what looked like a white sandy beach, with only the surrounding mountains breaking the impression that we were seeing something like an artificial tropical sea. This impression was misleading. Water only covered a tiny fraction of the dam (see figure 2). Similar to other tailings dams that we had visited, most of the surface was a monotonous light gray emptiness, totally flat and seemingly borderless, where nothing grew or lived. Kilometer after kilometer of lifeless gray.

Emptiness was accompanied by stillness. From our vantage point that day the dam looked like the epitome of tranquility. Nothing seemed to be moving in the gigantic infrastructure and its environs, especially the millions tons of tailings right in front of us. We experienced a sort of glacial time, tailings as a mass that was not only immobile but seemingly immovable, its

FIGURE 2. Satellite view of the Carén Dam

stillness difficult to imagine changing substantially in the short term, even in the decades to come. It was a gray world, seemingly firmly located in the deep time of geological processes.

The gray worlds of tailings are not so still, however. Tailings are described in the technical literature as "mixtures of crushed rock and processing fluids from mills, washeries or concentrators that remain after the extraction of economic metals, minerals, mineral fuels or coal from the mine resource."[3] Besides an indication that the term's first known usage was in 1764, little is known about the origin and early meanings of *tailings*.[4] This lack of clarity also results from the fact that historically several terms were used interchangeably to describe the same substance: in English *sludge, slag, slime*, and *slum*; in Spanish *relaves, colas, jales, barros*, and *lamas*.[5]

These shadowy origins also apply to the substance itself. Tailings can contain a wide array of components, starting from the highly variable composition of the geological strata from which the crushed rocks have been extracted. In addition, a wide array of reagents could be added to them in order to enhance the recovery of valued minerals, including "organic chemicals, cyanide, sulfuric acid, and other[s]."[6] Some of these components, such as cyanide, are well-known toxicants, whose negative effects on human and environmental well-being have been known for a long time. However, many kinds of tailings do not include these potentially toxic elements in high concentrations, making them far less potentially damaging. Tailings are quite variable entities, hence their exact chemical composition varies from one mine to the next, even in the same mine at different points in time. As the ores, reagents, and processing technologies change, so do tailings.

Current Teniente tailings are composed of around 45 percent water; the rest is a solid fraction formed of clay (21%), silt (68%), and sand (11%).[7] Regarding their chemical composition, these tailings are mostly a mixture of sulfide minerals such as pyrite, covellite, and bornite. Although not very toxic in themselves, their small particle size and their rapid dry-out in the semiarid conditions of the basin mean that these sulfide minerals could easily oxidize, generating acid mine drainage, one of the most common forms of pollution associated with large-scale mining.[8] Other minerals found in these tailings are copper, zinc, and molybdenum. Although their relatively low concentrations diminish their potential toxic effects, at least molybdenum has been of concern for both mining operatives and authorities in recent years (as we will see in chapter 2).

Besides their potential toxicity, what makes the issue of tailings' environmental impact pressing is that even midsize mines produce thousands tons of tailings *per day*, in all reaching a "global quantity of approximately 18 billion m^3 per year . . . [which could be on] the same order of magnitude as the actual sediment discharge to the oceans."[9] Traditionally, this enormous amount of waste was simply washed out to the nearest body of water, taking advantage of tailings' semiliquid condition. This practice caused massive damage, completely obliterating ecologies downstream. As a way to deal with this situation, since the beginning of the twentieth century most mines (although not all) have opted to build massive dams to store tailings.

The dams where tailings are accumulated usually become enormous, extending for hundreds of square kilometers and reaching depths of several hundred meters. Thus, it is not surprising that a survey on the topic called them "probably the largest man-made structures on Earth."[10] Given this scale, the whole process of producing and accumulating tailings (along with other mining waste products) not only "comprise[s] the world's largest industrial waste stream" but also is seen as on "the same order of magnitude as that of fundamental Earth-shaping geological processes."[11] For example, already by 2007, Embalse Carén "occupied an 8.5 km-long and up to 2.5 km-wide area of the valley. . . . With an average deposition of ca. 200,000 t/d tailings . . . , the total volume of the Carén dam was estimated at $7.6Mm^3$ with a total surface of 22 km^2."[12] What we were seeing that day at the dam wall, putting it brutally, were the very bowels of the Andes.

Given their potential toxicity and enormous volume, tailings have become the most important source of mining-related pollution and destruction. A cadastre on the issue carried out by the official industry organization, concluded that worldwide since the 1970s there has been at least one major failure of a dam each year.[13] The identified causes of these faults are multiple, including, among others, liquefaction under seismic activity, extreme pressure at the dam wall, ground faults, excessive water levels, and leaks. Such failures have caused massive ecological and social damage, becoming in some cases the worst environmental disasters experienced in the affected countries, as happened in the Bento Rodrigues (2015) and Brumadinho (2019) disasters in Brazil. Worryingly, such failures have increased substantially during the ongoing mining supercycle, especially due to the growing exploitation of complex ore bodies with lower concentrations of valued minerals, which involve the production of more tailings.[14] In addition to the risk of utter collapse, tailings ponds are associated with forms of regular pollution, as variable amounts of tailings regularly spill out of the dams, polluting local streams and being carried away by the wind as dust. Dams constantly break and leak, especially older ones.[15] Given their huge volumes and geomorphological properties, their multifarious risks and toxic potentialities, and their "deep time" implications, we could easily see tailings dams as a key materialization of the Anthropocene.

Like all things related to the Anthropocene, tailings and their destructive capacities are also crisscrossed by deep patterns of inequality and violence. Although tailings ponds are among the largest structures ever made, most people—especially in the affluent West—have never seen one, much less been affected by its contents. This is not accidental. The largest and most risky tailings dams worldwide are located in medium- to low-income countries, places already marked by high levels of social inequality, weak governance, and systemic corruption.[16] Inside such countries, the dams also tend to be located in areas of little value, usually far away from the centers of local power, places where other discards of global capitalism already reside: among the urban poor, neglected ecologies, indigenous communities, invasive species, and the like. Obviously, no member of these groups is ever consulted about the arrival of such a massive (and troublesome) neighbor. The locations of these dams therefore are not only a practical or economical matter, but reflect (and amplify) deep patterns of social and environmental inequality.

This forced cohabitation usually goes hand in hand with multiple forms of violence. Tailings transport, disposition, and (mis)management are usually accompanied by a series of malpractices typical of mining corporations.[17] Practices such as land grabbing, violence against communal leaders (especially indigenous populations), damage to farmlands and cultural heritage, and a long list of others continue to characterize an industry that historically "has taken a 'devil may care' attitude to the impacts of its operations."[18] It is not surprising that mining is usually seen as one of the most socially and environmentally destructive industries, especially when it is set in the colonial mode of transnational corporations operating in remote locations, where they encounter little regulation and accountability, as usually happens in Latin America. The few valuables that corporations actually extract from such locations—the high-grade minerals—rest on the creation of vast tailings flatlands, always prone to collapse and pollute, whose management usually involves adopting violent means to address any kind of local opposition.

Reacting to this violence, since the 2000s massive social movements emerged against mining throughout Latin America, some of them directly focused on the nefarious environmental effects of tailings and tailings dams. Under the concept of *extractivismo*, mining in Latin America is conceived of as an enterprise whose sole focus is the export of raw materials. As declared by Gudynas, "in extractivismos nothing is produced, there is only an extraction."[19] Extractivismo "refers to a 'mode of appropriation,' rather than a mode of production," appropriation that in most cases only benefits a small local capitalist elite and, increasingly, transnational corporations.[20] For most of the affected others, extractivismo is only a source of dispossession and poverty, as well as "serious and irreversible damage to the natural environment."[21] Derived from this conception, these movements are commonly focused not only on rebalancing the distribution of wealth and well-being, but on the utter elimination of mining projects. Increasingly organized as transnational networks, movements against extractivismo have rapidly become one of the most prominent forms of environmental activism in countries such as Chile.

Mineral extraction is damage and pollution, it seems, especially when seen from the optics of tailings, the monstrous "tails" of capitalist extractive efforts. These tails are so massive that not only damage local people and ecologies on a daily basis but are even reconfiguring the planet's geomorphology. In multiple locations throughout the world, tailings dams actualize colonial regimes of power and violence, cementing over all kinds of local life projects, especially the ones already affected by multiple other forms of precariousness.

FIGURE 3. Downstream view from the Carén Dam wall

A GREEN PLACE

Upon reaching the other side of the wall that day, our initial assessment of Carén Dam was rapidly challenged. Besides the wall itself, the view downstream clearly departed from the image of damage and pollution we associated with tailings ponds. A narrow creek emerged at the bottom of the wall and then ran down towards the valley. Although crossing through bare lands at first, a bit farther down it was surrounded by a fair amount of greenery, both agricultural fields and forests. The view downstream (see figure 3), appeared to be an open negation of what we had seen on the other side of the wall.

This impression only confirmed what we had seen upon approaching the Carén complex after a two-hour drive southwest from Santiago. From the road, we didn't see any signs of the million tons of potentially toxic tailings accumulating just a few kilometers away. There was no gray wasteland, no massive pieces of mining-related infrastructure or the dark-orange-colored water associated with acid mine drainage, as we had observed in the vicinity of other tailings dams in Chile. To our eyes, the basin appeared as fairly typical of Chile's central valley, relatively arid but with plenty of agricultural fields, with crops such as corn and potatoes, plus livestock, interspersed with some houses and a couple of small businesses. After crossing the gates of the Carén complex, we saw not only a highly productive experimental farm but even a beautiful lagoon surrounded by grown trees, home of several wild bird species.

Such a view was even more surprising considering the so-called megadrought affecting central Chile.[22] Lasting more than ten years, this event shattered all the recorded minimum levels of water precipitation, being widely seen as the foremost manifestation of climate change in the country. The extreme lack of water has progressively turned areas in the direct vicinity of the Carén basin into drylands. If agriculture is still possible there, it is only because massive infrastructure has been erected to extract groundwater, usually involving wells hundreds of meters deep. As most small landowners cannot afford the investment needed to drill such wells, they have been forced to sell their lands to large agro-industrial corporations, Chile's other main extractive industry.

The name Carén comes from a combination of the Mapudungun words *karv* (green) and *we* (place). And this was largely what we had found there since first arriving to do fieldwork: a relatively green place in the midst of the anthropogenic devastation caused by the megadrought. This greenery was a product not only of the presence of deep wells and large corporations, but of dozens of low-income landowners cultivating relatively small plots of land, usually in a highly artisanal fashion. In a way, to the visitor the basin appeared as a relic from the past, one in which surface water was still regularly available, so with little investment one could run a productive farm. And this was *because of*, not *despite*, the tailings dam.

Carén, we rapidly realized, was a paradox.

This paradoxical character started from the very top. CODELCO was created in 1976 as a publicly owned company in charge of the management of the recently nationalized copper mining industry (further explored in chapter 1), becoming as a result the biggest copper mining corporation in the world. In achieving this status, its operation has been repeatedly associated with most of the conventional negative effects of large-scale industrial mining in the global south: environmental degradation; violence against local communities, especially indigenous people; destruction of cultural heritage; and so on.

At the same time, CODELCO does several things that are much less common for large mining corporations. First, and foremost, being a publicly owned company, its revenues do not go to some group of wealthy individuals located in a faraway country. On the contrary, they go directly to the Chilean state and are in fact its primary source of revenue. Second, CODELCO is expected to act in a state-like manner, considering the public well-being along with the search for profits. Even before the current hype about corporate social responsibility (CSR) in the mining sector, CODELCO was for years aiming at achieving higher standards in social and environmental issues than other, privately owned, mining corporations operating in Chile. Although these initiatives do not usually go beyond establishing clientelistic relationships with communities surrounding the mines, they raise not inconsiderable points of friction with the mine's daily operations.

These contradictions have been very much present in the case of Teniente. Since its very beginning, the mine has focused not only on the production of copper, but also on the management of local populations. This aim was first directed toward turning them into trustable and loyal workers.[23] As critical voices were raised against the mine's environmental impact beginning in the 1990s, the program was progressively enlarged to include the population living in the mine's vicinity, establishing clientelistic patterns of relations with them (usually based on the provision of various goods and services). As we experienced frequently during fieldwork, CODELCO's behavior and image among locals were clearly different not only from the image of mining corporations elsewhere in Latin America, but even from analyses of the operation of CODELCO at other locations in Chile.[24] In line with recent ethnographic explorations of mining corporations, the version of CODELCO emerging from our fieldwork at Teniente was many things at once.[25] Certainly it was capable of embodying the monstrous corporation of extractivismo narratives, but it also was able to become a force locally recognized as having a positive impact, even a source of pride, the "orgullo de todos" (pride of everyone), as its slogan declares.

Finally, there was the issue of water. As mentioned, the Carén complex is located 85 kilometers east of Teniente, across Chile's central valley. To bring tailings there, it was necessary to build a massive canal, which includes several large bridges and tunnels and passes through areas of intensive agricultural activity and urban centers. Tailings move solely by the force of gravity through this canal, so it is necessary to add a significant amount of water to help them drain. In most mines in Chile this water is afterward returned to the processing plant. In the case of Teniente, however, the distance and geographical obstacles make this return impractical, so the water that comes with tailings is released in Carén Creek after being treated. At Teniente extraction involves mobilizing not only ores and tailings, but also water, turning the dam into a lifeline for the whole basin, a paradoxical antidote against the megadrought.

We had arrived at Carén expecting to find extensive pollution and obliterated ecologies. We certainly found them, as the chapters of this book testify. But we also found many other things, several of them openly contradicting the destruction usually associated with mining environments. There were productive agricultural fields. There were strange wildlife entanglements. There were people who saw this as a "natural" spot. There was damage, for sure. But also, there was life. *Life within extraction.*

When trying to decide how we should read the material collected there, these multiple encounters between tailings and life, we found ourselves increasingly at odds with most social science literature on the extractive industries, especially in Latin America. Against critical notions on the extractivismo literature of mining as the "*singular point of origin* of a range of social, economic, and environmental pathologies," in Carén we found tailings causing both disruption *and* emergence of life, usually in the same movements.[26] The pathologies were there, but usually accompanied by several unexpected vital developments, some of them even pointing toward tentatively hopeful futures, to less-damaging, (even) fulfilling ways to live with tailings and other residues of industrial processes.

In this book we aim to delve into Carén's contradictions to explore the possibility of thinking (and doing) extraction differently. Following Donna Haraway and many others in the environmental humanities and science and technology studies (STS), we aim at "staying with the trouble" of tailings.[27] Staying with the trouble of tailings necessarily implies properly *seeing* them, assigning them analytical and ethnopolitical space, and recognizing in them a certain dignity, even a right to existence.

GEOSYMBIOSIS

A first hint of a more generative approach to Carén was suggested when we read a geochemical description of the tailings produced by Teniente:

> Mill tailings, derived from mining sulfide-bearing ore deposits, are essentially composed of crushed rock. These systems are typically devoid of organic carbon, which limits biological cycling of sulfur, iron, and other metals in these environments. Sulfide-oxidizing bacteria, such as Acidithiobacillus

ferrooxidans, are autotrophic and can assimilate organic carbon from CO^2. These bacteria derive energy through the catalysis of inorganic reactions. Microorganisms can act directly in the oxidation of metal sulfides . . . as a catalyst for ferrous to ferric iron oxidation . . . , in element liberation, as well as in retention and neutralization processes in mine-waste environments. . . . The interactions between the mineralogy, microorganisms, and organic metabolites are some of the key parameters to understand the formation of contaminated mine waters.[28]

Tailings are not dead. From their very first moments, as they travel the canal toward Carén, tailings become home to multiple kinds of microorganisms, especially bacteria such as *Acidithiobacillus ferrooxidans*, the main agent in their oxidizing. Although the existence and capacities of such bacteria have been known by the industry for awhile, generating a whole set of technical processes to use them to speed up the extraction process, it is also clear that "the metabolic capacities of microbes have their own rhythms . . . [and] transcend spatiotemporal limits associated with extraction of metals from low-grade and complex sulfidic ore, they produce new ones."[29] Bacteria produce new orderings in the dam, as they become entangled with a plethora of organic and inorganic entities. In doing so, bacteria—along with many other organic entities, as we will see—utterly transform the dam from a place of gray death, the "ultimate sink" for the industry's residues, into a place teeming with life.[30]

The presence of bacteria in tailings directed us to the work on symbiosis carried out by multiple evolutionary biologists since the end of the nineteenth century, most notably Lynn Margulis, and their forays into the environmental humanities and STS.[31] Breaking with conventional readings of Charles Darwin's evolutionary theory as pure competition, these researchers posited the need to understand the evolution of life on Earth also as a matter of symbiotic entanglement, especially at the bacterial level. By symbiosis Margulis simply meant "the physical connection between organisms of different species."[32] The evolutionary relevance of this process is that long-term contact could produce *symbiogenesis*, or "the appearance of new bodies, new organs, new species," resulting in more complex life forms.[33]

Given the prominence of such symbiotic entanglements in most ecologies, the very concept of individuality loses meaning. Any organism, especially complex ones such as human beings, should be seen instead as *holobiomes* or as "a collection of interpenetrating ecosystems," formed by a vast array of smaller organisms entangled in symbiotic relationships.[34] Any kind of development—human or otherwise—implies "becom[ing] with others" in multispecies entanglements; "symbionts and hosts do not lead independent existences.[35] Rather, they are the mutual cause of the other's development."[36] Instead of being merely the background for a well-defined set of individual species, an ecology is growingly seen as an ever-changing arrangement that "organisms actively modify to suit their responses" but that modifies them in response.[37]

In this process of continual symbiotic emergence, conventional nonliving entities such as minerals occupy a central place. Not only do minerals affect biological entities, forming a major part of their structure, but biological entities "are intimately associated with the biogeochemical cycling of metals."[38] For Caldwell and Caldwell, this process leads to the emergence of what they have called *geosymbiosis* or "a reciprocal relationship in which the restructuring and proliferation of a mineral affects the proliferation rate of an organism, and the restructuring and proliferation of the organism affects the proliferation rate of the mineral."[39] In a similar way to purely biological symbioses, "both geosymbiotic partners affect and are affected in return," representing "cross-linkages between the pipelines of biological and geological innovation" and even leading to the formation of whole new holo(geo) biomes.[40] Some of these geosymbioses can lead to a flourishing of biological life in places with intensive concentrations of heavy metals, such as in phytoremediation. In other cases, such as acid mine drainage, geosymbioses can negatively affect several biological entities. What is undeniable is that minerals are not passive recipients of symbioses but very much participants in them, being importantly changed as a result. Even conventionally "toxic" chemical compounds could be seen as "[geo]symbiotic partners of coevolution," not necessarily attacking biological entities but taking them somewhere else.[41]

To recognize that a certain degree of entanglement is at least as important as competition for the emergence and maintenance of life does not mean that symbioses are necessarily an improvement for all the entities involved, as is usually assumed. As stated by Sapp, one of the more complex issues regarding the historical development of symbiosis theory has been the tendency among many of its practitioners to equate symbiotic entanglements with social and/or moral orderings.[42] Against the rampant spread of global capitalism and its unabashed support of individual competition, symbiosis appeared to "naturalize" an alternative social ordering in which cooperation and mutual care was the rule, not an exception. Such a tendency to moralize symbiosis has been strengthened through its uptake in the environmental humanities.[43]

However, in practice "this narrow meaning of the term is difficult, if not impossible to apply to real associations," given that many symbiotic entanglements are anything but mutually beneficial or caring.[44] In this regard, the moralization of symbiosis could be seen mostly as an example of what Daston calls a "naturalistic fallacy," or the practice of transferring human values to nature to give them greater authority.[45]

How can we engage with geosymbiosis while avoiding the trap of moralization? In our case, by adding a further component to conventional models of symbiosis. In accordance with such models, symbioses range from mutually beneficial (usually known as mutualistic) or neutral (known as commensal) relationships to detrimental interactions (known as parasitic). Parasitism, in particular, is conventionally seen as an unequal relationship in which "the symbiont is using the host as a resource and causing it harm as a result."[46] However, such harm can be tolerated by the host "either because the parasite to some degree restrains its attack upon the host, or because the harm can be absorbed or compensated for in some way by the host species."[47]

There are situations, however, in which the harm caused by the symbiont cannot be tolerated without suffering great damage, damage that can even lead to the death of the host.[48] Regarding the entanglements between biological and mineral entities, we call these situations *toxic geosymbioses*.

From an environmental toxicology perspective, a toxicant is "a substance that occurs in the environment ... and has a deleterious effect on living organisms."[49] For such deleterious effects to occur, some form of interaction must happen between the entities. Toxicity is then mostly understood as "the manifestation of an interaction between molecules constituting some form of life and molecules of exogenous chemicals or forms of life affected by physical insults."[50] Toxicity is not an "external" attack on life, as usually portrayed in warfare-like models of disease and purity, but something more akin to an intimate relationship, even a "chemical kinship."[51] The fact that toxicity is intimate, however, does not mean it is solely located in bodies. As a growing literature has explored, most toxic entanglements nowadays are consequences of global regimes of colonialism, racism, violence, and exploitation.[52] They are molecular manifestations of the highly unequal distribution of benefits and harms characteristic of contemporary capitalism, as we explore throughout this book.

Toxicity is therefore seen here as a particular kind of geosymbiosis, an entanglement between biological entities and mineral compounds that damages in an intolerable way the former, leading to permanent harm, even its death. This entanglement is still symbiotic because generally it produces some marginal benefits to the biological host, some temporary relief from other needs, or the paradoxical freedom resulting from being too toxic to be eaten or controlled.[53] These benefits, however, are temporary, as toxic geosymbioses always lead to the ultimate ruination of the biological host. These entanglements are always relatively temporary, as they last only as long as the biological host can resist the damage caused by the chemical geosymbiont. They either are turned into less-damaging forms of symbiosis, especially parasitism, or lead to death.

When toxicity is integrated into the conventional model of symbiosis, as shown in figure 4, there is certainly a rebalancing. No longer are geosymbioses seen as inherently good or moral, because as much as they can cause mutual benefit, they can also lead to the utter obliteration of one of the partners. This inclusion also highlights other relevant characteristics of geosymbioses. First, geosymbioses are usually unequal in terms of costs and benefits. While the chemical components frequently are able to proliferate through them, the proliferation of the biological partner is less certain. In many cases, the chemicals act as exploitative symbionts, proliferating as much as possible without taking into consideration at all the well-being of the biological host.[54] As has been beautifully summarized by Povinelli, most of the time "humans [and other living beings] are merely a moment on the journey and travels of minerals."[55] Second, geosymbioses have a relatively shorter temporality. On the one hand, while biological symbioses could take centuries to emerge, geosymbioses tend to be much faster, being able to emerge within the lifetime of an individual. On the other hand, especially in the case of toxic symbioses, they also tend to last for a shorter period, as the biological host will either sever the unequal interchange if it causes too much damage or perish by it. In this regard, geosymbioses are continually emerging and perishing. Third, geosymbioses are commonly interrelated with sociopolitical issues. Especially in a world crisscrossed by multiple chemical compounds resulting from human actions, many geosymbioses not only happen because of environmental conditions but also are intimately connected with issues of power, inequality, and violence. Or in other words, the movement from mutualistic to toxic geosymbioses usually matches unequal distributions of power characteristic of current capitalist societies. As we explore through the concept of residualism, many of the toxic geosymbioses found at Carén can be directly connected to the political economy of contemporary extractive industries and its intimate interlocking with multiple forms of socio-environmental violence.

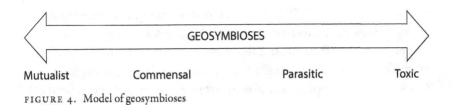

GEOSYMBIOSES

Mutualist   Commensal   Parasitic   Toxic

FIGURE 4. Model of geosymbioses

From this perspective, a different image of the Carén Creek basin emerges. Rather than being seen as something akin to mineral pathogens, tailings appear as able to nurture multiple geosymbioses, first between minerals and bacteria on the microbial level, but rapidly including other—much larger—beings, such as plants, farm animals, and humans. Some of these geosymbioses are toxic, no doubt about that. But many others are not, creating the basis for an extensive array of living arrangements in the dam and its environs.

BETWEEN BIOS AND GEOS

Exploring Carén from a geosymbiotic perspective locates our book on the threshold of two of the most important intellectual projects emerging from the environmental humanities and STS in recent decades. At the center of both there is a shared criticism of the anthropocentrism characterizing the modern episteme, seeing it as a leading cause of our current ecological troubles, from climate change to species extinctions. They both explore alternatives to anthropocentrism, putting the focus on the nonhuman world as a way to emphasize humanity's interconnection and dependence on processes and entities existing well beyond human beings.

The first such project has placed its primary focus on the *nonhuman bios*, or the myriad nonhuman biological entities populating our world. Known as multispecies studies, it understands organisms, both human and animal, as always "situated within deep histories of relatedness in which . . . being is always becoming-with."[56] The world emerges out of these relationships, fluctuating and multivariate, a world in which highly heterogeneous entities actively participate, none of them being in control all the time. In line with symbiosis theory, human beings emerge "through shifting, usually asymmetrical, relationships with other beings."[57]

Given this framework, the primary focus of multispecies studies becomes "to study the myriad organisms whose lives and deaths are intimately linked to human social worlds."[58] In doing so, it seeks not only to develop a new in-depth look at the interrelationships between humans and nonhumans that produce our lifeworlds, but also to intervene in them, engaging in forms of writing and action that might help them to survive in challenging times. Contrary to catastrophist narratives of utter environmental and societal collapse, this analytical/practical immersion in the interrelationships between humans and nonhumans is intersected by forms of precarious hope, not for a "return to nature" or for the realization of "sustainable development" but for the emergence of new forms of mutual care and attention that could allow for the survival of (at least) some spaces for multispecies flourishing in the midst of planetary devastation.

The second project puts its main focus on the *nonhuman geos*, or the multiple chemical substances and geological processes shaping life on Earth. In contrast to the multispecies program, the main focus of this literature is "not only how to extend or enrich the composition of shared worlds but what to make of forces capable of interrupting, undermining or overwhelming the very conditions of doing politics or being social."[59] At the center of this program is the ultimate indifference of the nonliving world toward humanity, an indifference that points to a radical asymmetry, because while we are very much dependent on these forces to continue existing, most geological entities and processes would be mostly unaffected by our disappearance.[60] So, a main task of critical thought becomes to "imagine new political geologies as an inhuman agency that is not and cannot be fully co-extensive with the human domain, however inclusively this is imagined."[61] Such a politics should start, following Povinelli, by challenging current regimes of *geontopower*, or a tactic of power that governs through the strategic management of the distinction between biological life and geological nonlife, at the heart of most colonial regimes.[62] What is critical for Povinelli is that commonly even notions of interconnectedness such as those mobilized by multispecies studies appear as "reiterating rather than challenging the discourse and strategy of geontopower."[63] In contrast, these authors aim at recognizing "the geologic as a praxis of differentiated planetary inhabitation and corporeal affiliation, rather than an externality."[64] This "geological life" confronts humanity with scales and temporalities that are radically out of reach, forcing on us no little degree of passivity and humility, the recognition that not everything is up for human appropriation and transformation, the great paradox of the Anthropocene.

Certainly we are not the first to aim at connecting both intellectual projects. There have been several efforts exploring pathways between the geos, including anthropogenic chemicals, and the bios. Their first move is the recognition of the impossibility of reestablishing former limits separating the two. We are living in a time in which novel mixtures of the bios and geos are happening everywhere, all the time, especially due to human intervention. Given this, most demands and/or searches for purity appear more as a tactic of power rather than an achievable goal.[65] The aforementioned efforts voice urgent calls for devising novel ways to live among the ruins, especially regarding less-damaging, more just, and even tentatively satisfying ways to develop the human and nonhuman bios in our contemporary "chemospheres."[66]

Increasingly this move has been paired with a break from the damage-centered narratives that tend to characterize research on the topic, narratives that usually "essentialize those who bear the disproportionate burdens of harm as victims," with the consequence of "rendering lives and their landscapes pathological."[67] To maintain distance from damage-centered narratives does not mean to deny the existence of harm or the nihilistic celebration of suffering and decay, in the form of an environmental ruin porn. Damage persists, for sure, and on scales unseen before. And such damage is unequally distributed, a leading manifestation of ongoing colonial regimes of capitalistic exploitation. But there is also desire: "Desire, yes, accounts for the loss and despair, but also the hope, the visions, the wisdom of lived lives and communities."[68] Taking full account of the constant damage and violence caused by manmade intromissions of the geos into the bios, and the pessimistic acknowledgment about its inescapability, these narratives explore stories of resistance, resurgence, and transformation, tentative paths forward from the death politics of the Anthropocene.

This approach has also been taken by recent research on extraction in Latin America.[69] This literature has been inspired by two interrelated swathes of decolonial thought. On the one hand is Latin American ecofeminist thought focusing on principles of the "circulation of life," or the search for alternatives to extractivism based on a defense of more equalitarian relations between genders and beings.[70] On the other hand are authors analyzing Amerindian ontologies as a way to reimagine the bios and the geos as nonhuman beings with their own world-making capabilities and rights.[71] In all, this literature has sought to reveal "a differently perceivable world, an intangible space of emergence, where rivers converge into the flow and the muck of life otherwise."[72] From this focus, even in the devastation caused by extraction there are possibilities for the reemergence of life. Such reemergence does not always take the form of open resistance, but also adopts the very materials and strategies of extractivist projects, deploying them in twisted, subversive ways.

In this book we aim at contributing to this field. Through our focus on geo-symbioses at Carén we intend to depart from a conventionally negative view of tailings as inert waste, solely a source of pollution and destruction.[73] Tailings at Carén, like any other chemical entity, entangle in multiple ways with other entities. Some of these entanglements are toxic, even deadly. Others are vital, allowing the proliferation of multiple forms of life. Most of them, we suggest, exist between these extremes, allowing *certain* types of life in *certain* ways, but canceling out many others. In taking such an approach we also aim at extracting the entities affected by the interactions—from algae to human beings—from purely damage-centered narratives. To be clear, by doing this we do not aim to normalize or hide damage. Nothing of the sort. Our aim is simply to show how damage goes hand in hand with desire, with the pulsion among the affected entities to continue living, to persist and proliferate, even if such living entails establishing alliances with the very substances causing the damage.

At the same time, and closer to our STS sensibilities, this is also a story of the weakness of technoscientific power in the Anthropocene. By putting a dedicated focus on Teniente managers and technical personnel and their troubles in keeping geosymbioses at bay, we aim to highlight the limited degree to which massive anthropogenic entities such as tailings dams can be managed, even less in a sustainable way. Like Kolbert, what we saw in Carén was "not so much the control of nature as the *control of* the control of nature," or attempts to govern the overflows caused by previous failed attempts to control tailings.[74] And such attempts were, most of the time, only partially successful, if at all. In line with the geos literature, our story aims to show that we, as humans, have little control over the multiple geosymbiotic processes happening inside and outside places of extraction, that besides the original act of deciding to put tailings in that particular basin, humans have largely surrendered its control to countless entanglements between biological and mineral entities, enacting what we have called *symbiopower*. Carén is alive, as one of our interviewees told us, but not alive *for* us. It is alive to proliferate, even if such proliferation (as commonly happens) goes against human and animal survival.

In exploring tailings' symbiopower and the weaknesses of control, we tried to set aside our own previous judgments, forcing ourselves to simply see what emerges from tailings, which kind of life they nurture (and destroy), and how little control human beings have over what happens. In doing so, our book aims to contribute to the urgent task of "hold[ing] open space for possible multispecies futures in the midst of escalating extractions and extinctions."[75] Along with the urgent task of telling stories of extinction, we also need to tell stories of the emergence of new ecologies in the Anthropocene, in our case, the monstrous geosymbioses of tailings. In doing so, we aim to see them as entanglements with a certain dignity, Frankensteins always unloved, but whose voice is more relevant than ever. Because we—human, animal, mineral—*entangle with* them. Once we live in an area on which million tons of tailings have accumulated, we *entangle with* tailings. Mining becomes biology; extraction becomes evolution.

A PATCHY ETHNOGRAPHY

As Ballard and Banks recognized in their overview of the field, the mining industry is "no ethnographic playground."[76] Given the amount of socio-environmental conflict and controversy surrounding the industry everywhere and the complex position of social scientists as both critics *and* facilitators of it, in doing an ethnographic study of mining we are also entering "a parallel war of sorts . . . about the nature and scope of appropriate forms of engagement."[77] For academic researchers, this conundrum has usually led to the tactic of "'[keeping] their distance' from the company they studied and [holding] it 'at arm's length,' emphasizing their solidarity with those who stood outside company fences."[78] As is already evident, we have not followed such a strategy. In line with other recent ethnographies of mining, we have opted for "staying with the trouble" not only of tailings, but also of a complex and contradictory mining corporation such as CODELCO.[79]

Our book is based on roughly three years of analytic engagement with Teniente, especially at Carén Dam, from 2013 to 2016. This engagement consisted, first, in accompanying CODELCO personnel and external consultants on their daily duties in and around the dam. This initial approach had the advantage of providing us with a very intimate view of the practical and ethical dilemmas of dealing with this massive and unruly entity, the "dragon," as one manager described it. However, this approach also had some costs. Differing from other ethnographies of mining, the core of the collected material was not derived from a continual residence in the area. Each access to Teniente facilities was granted only after lengthy negotiations and for only a few days at a time, a situation that forced us to establish an irregular pattern of visits, spending only a few days inside the mine's premises each time.

These limitations, however, had the positive effect of forcing us to look elsewhere. First of all, we regularly engaged with locals, especially with farmers living downstream from the dam, spending many days in their fields, talking to them and watching how they dealt with CODELCO, the dam, and especially, the water flowing through Carén Creek. All this work was complemented with extensive archive material collection, ranging from CODELCO official brochures to judicial documents. These heterogeneous tactics resulted in what we term a *patchy ethnography*, or a situated, irregular, hybrid form of doing research and narration, in which contemporary accounts freely mix with historical ones, a focus on human beings with a multispecies sensibility, description with speculation. After all, "to write a history of ruin, we need to follow broken bits of many stories and to move in and out of many patches."[80] This is a kind of research, we hope, especially well suited to engaging with our patchy Anthropocene.[81]

We want to finish with a note of caution. We are fully aware that this exercise could be easily misinterpreted as a new way to glamorize and/or normalize destruction, geosymbiosis becoming a sort of CSR 2.0 for an industry always thirsty for cost-effective techno-fixes, as already happened to us in the process of getting access to the field.[82] No matter how much we emphasize that the anthropogenic geosymbioses we studied are not "nice," not even remotely comparable with the ecologies that evolved for thousand years on these locations, the risk of misinterpretation will always be there, forcing us to be especially vigilant about the ontological politics of our book or the realities it helps to nurture or inhibit.[83]

For this reason, we want to strongly emphasize the exceptionality of our case study. As the then director of Chile's Comisión Nacional del Medio Ambiente (CONAMA) recognized in 2006 in the quote that opens this introduction, "there is no other tailings dam in Chile with the characteristics of Carén."[84] Carén is a rarity, an exception.[85] Even to people in the industry it appears as an anomaly, a monstrosity, a tailings dam that nurtures life of many kinds, not only through its "ruins," but in its daily operation, *within extraction*.[86] After reviewing a great deal of literature on mining, both technical and from the social sciences, we have found no similar case, not in Chile or elsewhere.

Through studying Carén's exceptionality, we hope to demonstrate that environmental devastation *is not necessarily* inherent to extraction. Although no little amount of damage is compulsory to extractive operations, this does not equate with devastation or the utter abolishment of most possibilities of life in the areas where extractive industries operate. If such devastation occurs, as daily happens in mines throughout the world, especially in the global south, it is mostly because the actors involved—from companies to authorities, the media, and even the communities affected—*do not care enough* to prevent it. They are not willing to intervene and/or capable of intervening by nurturing more mutually beneficial entanglements and avoiding toxic geosymbioses. The reasons given for such a lack of care are manifold—financial, political, cultural, environmental, and so on—but they all point to a consideration of possible geosymbioses as unimportant, as merely poor replacements for "real" nature and hence not worth fighting for. Through analyzing tailings and their imperfect geosymbioses in Carén, through enhancing their vitality and dignity, this book aims to disrupt this narrative and offer tentative hope, lifelines crisscrossing our multiple devastations.

STRUCTURE OF THE BOOK

The book consists of six chapters. In order to set the scene for a proper engagement with current geosymbioses, chapter 1 relates a genealogy of the tailings at Carén, asking how they came into existence in the first place. Two key historical developments are identified in this emergence. First is the development of flotation technology for mineral separation at Broken Hill, Australia, in the first decades of the twentieth century. Flotation was the process through which tailings originally emerged, as massive residues with a loose geochemical structure. Second is the gamble taken by Teniente's then owner in bringing flotation to the mine a few years later, turning an almost abandoned project into one of the most productive mines in the world. In both operations we see the emergence of *residualism*, or a particular logic of production in which, on the one hand, the production of valued minerals depends on the creation of massive amounts of residues, which, on the other hand, are deemed invisible as both material and matters of concern. Residualism, the molecular project of mining capitalism, sees tailings as devoid of life, passive substances only activated by further waves of value extraction.

In geological terms, residualism turns into sedimentation or the belief that upon reaching Carén, tailings rapidly sediment into stable geological strata, hence ceasing to have any kind of potentially troublesome vitality. As explored in chapter 2, tailings at Carén rapidly disrupt this narrative after being deposited, engaging in multiple geosymbiotic processes inside the dam. At the organic level, the water emerging out of accumulated tailings becomes home for entities such as fish and algae. Given the practical impossibility of eradicating them, the dam personnel have been forced to create ways to cohabit with them somewhat, with varying degrees of success. At the inorganic level, due to their loose geochemical structure and huge volume, tailings are regularly affected by massive displacements in a matter of minutes, always risking a massive spill. Contrary to residualism, the main way CODELCO's personnel have dealt with these events has been to treat the dam as a living entity, a "dragon" that they cannot fully understand, depending mostly on its goodwill to avoid environmental disaster.

Chapter 3 deals with the biological program of residualism, the thesis that a happy coexistence between tailings and agriculture at Carén is possible. Given CODELCO's desire to demonstrate that it practices sustainable mining, since tailings started to arrive at the dam the corporation has been involved in elaborate attempts to depict the water extracted from tailings as innocuous, hence demonstrating that mining can easily coexist with agriculture in the basin. The main vehicle for mobilizing this argument was the creation of an experimental farm to scientifically test the effects of this water intake on farm animals and plants. This farm has sought to produce a rare kind of agriculture, one in which potentially toxic chemicals and biological entities such as crops and livestock are seen as closely coexisting *but* not really engaging in geosymbioses. The results of decades of experimentation at the farm—on which fertility is only possible through a certain amount of toxic geosymbiosis—shows again the ultimate failure of residualism, signaling the need to develop more comprehensive engagements with ecologies such as Carén.

Regardless of the research being done at the experimental farm, after being deposited in Carén Creek, the water coming from the dam has become a crucial component for multiple ecologies emerging downstream, as explored in chapter 4. First, the chapter analyzes the case of local farmers engaged in what we call a *parasitic agriculture*, or a form of agriculture especially adapted to Carén, challenging in the process legal and sanitary dispositions. Second, the chapter follows families picnicking on the creekbanks on weekends, seeing the act of swimming in the creek on a hot summer day as a leisurely appropriation of space that challenges conventional notions about healthiness and the enjoyment of nature. In all, we see both farmers and vacationers as unknowingly engaged in diverging forms of *parasitism*; they resist not by opposing or refusing tailings and/or CODELCO, but by engaging surreptitiously with them, even embodying them in unexpected ways.

In chapter 5 we explore the aftermath of the massive tailings spill that occurred in Carén in April 2006. To begin with, we analyze the spill itself, seeing it as a particularly destructive manifestation of the agential capacities of tailings, a vivid demonstration of how human control over geosymbioses is always partial and fragmentary. Then we explore the two responses deployed to deal with the damage caused by the spill: remediation and compensation. In the first case, remediation aimed at establishing a clear framework of action regarding the spill, a frame that was based on ecological purity, hence making automatically invisible most of the geosymbioses discussed in previous chapters. In the second case, the search for economic compensation through a lawsuit presented (and won) by a local farmer against CODELCO enacted the damage only as a matter of profits lost, negating any recognition to the suffering experienced by the affected nonhumans. Contrary to naive celebrations of resilience, both paths show us that the survival of life in areas such as Carén needs not just geobiological exuberance, but also the development of a new kind of politics, especially for extractive projects such as Teniente.

Beginning with a visit to Carén in January 2021, chapter 6 acknowledges that the conditions of possibility for most geosymbioses found there are, at best, tenuous. In this regard, the book finishes by presenting a proposal for a novel kind of relationship between anthropogenic compounds and human beings. It starts by recognizing the existence of a specific kind of power in symbiosis—what we have called, rather unimaginatively, *symbiopower*—referring to living and nonliving beings' capacities to entangle with entities different from them, even forming new species as a result. Symbiopower emerges well before/below any kind of human intentionality and control, but in turn, it has the capacity to affect us deeply, even menacing our existence. For this reason, in relating to organic and inorganic entities we have to develop a politics of weakness, a paradoxical kind of power that rests on our recognition that there are many processes in the world that we can only barely understand, much less predict or control. Enacting such politics would start by properly *seeing* tailings as products of mining, not merely residues, and carefully engaging *with* them in parasitic ways in order to enhance the more fruitful geosymbioses, while being attuned to the more dangerous ones. In all, a politics of weakness aims to offer a tentative path forward toward assuming the practical and ethical challenges inherent in humanity's newly discovered earth-shaping powers.

ONE

Residualism

COPPER MODERNITIES

At around 1:00 p.m. on November 17, 1986, the first-ever cubic meters of tailings reached Carén. This event marked the culmination of a project started in the late 1970s that aimed to provide a new tailings dam by which Teniente could project its operations well into the future. At an estimated cost of U$150 million, and involving not only the construction of the dam itself but also of a 90-kilometer canal to transport tailings there, Carén was one of the most expensive and complex projects in Teniente's history. The efforts and expense were seen as worth it because the corporation obtained as a result "an infrastructure with a high degree of safety and operational efficiency, low maintenance costs and ample capacity for the future," as claimed a few days later by *Semanario El Teniente*, the mine's official newspaper.[1] Carén was not only a waste depository, but a lifeline, not only for Teniente, but for CODELCO as a whole.

Although officially created in 1976, CODELCO cannot be understood without referring to the copper nationalization process carried out under Salvador Allende's socialist government (1970–1973). One of the central policies of his government, nationalization aimed, in the words of one of its leading architects, "to recover for Chile its copper wealth, in defense of the country's independence and political and economic freedom."[2] Nationalization looked not only to increase state revenues, but also to sever historical ties of dependence with foreign companies and governments. A nationalized copper industry was seen as the material backbone of a new kind of modernity for the country, one in which strong public corporations such as CODELCO would occupy the leading position in an extensive process of sovereign industrial development. In doing so, the corporation would provide not only revenues for social policies, but a sense of national identity and pride for Chileans. The largest mine comprising CODELCO, Teniente occupied from the very beginning a central position in this process.[3]

As might be expected, this copper modernity (a national development Project built around copper extraction) was severely impacted by the coup d'état of September 11, 1973. Besides its terrible record regarding human rights violations, the regime headed by General Augusto Pinochet is known for carrying out one of the world's earliest and certainly most extensive processes of neoliberal public reform. The neoliberal program included not only the economization of most aspects of state action, enacting markets as the ultimate public space, but also a massive process of privatization of public assets, especially public corporations. Since it was the largest corporation in the government's hands, the "privatization of CODELCO was from the beginning a central goal . . . [for the country's novel] techno-bureaucracy."[4]

Given the military authorities' open resistance to CODELCO's outright privatization, the strategy deployed by its promoters within the government was to present it as utterly inefficient, so the military would finally be forced to accept that the state could not carry such a burden and allow (at least) some degree of privatization.[5] With this aim in mind, first, almost all the existing barriers to private corporations' exploiting the country's subsoil were removed. The subsequent arrival of large transnational corporations put a lot of pressure on CODELCO to remain competitive, not only internationally but now also locally. In addition, the corporation was required to significantly increase production while at the same time a "rationalization" of the corporation's budget was enacted, so that until the early 1980s "CODELCO barely received enough funds to maintain its existing production capacity."[6] Given this situation, enacting any large-scale investment project at the time was anything but easy. On the contrary, it had to be fiercely fought "through a combination of technical expertise, cooptation of kin agencies, and secrecy."[7]

The arrival of tailings at Carén that day in 1986 was a triumph. On the one hand, it was a demonstration that the corporation could enact complex new infrastructural projects under the harsh financial conditions imposed by the new authorities. On the other hand, the availability of a massive new tailings dam, one that was expected to remain operational until 2064, allowed the mine to significantly increase production in the subsequent years. As lyrically celebrated by *Semanario El Teniente*, through Carén "the men of Teniente ... [were able to] twist nature's hand," simultaneously twisting the *invisible hand* of the neoliberal market economy.[8] Out of infrastructure such as Carén a new copper modernity emerged, one in which CODELCO has remained "the biggest copper producer in the world ... [while meeting] the competitive pressures from the private sector and simultaneously play[ing] a more political role due to the strategic nature of extractive industries."[9]

In recent years, a further turn in this narrative has emerged: sustainability. In the wake of the ongoing transition toward renewable energy sources demanded by climate change—with Chile a world leader in solar energy production—copper has been regularly presented as a crucial component of such a process. In a context of growing social opposition against mining, this imaginary of copper as a key "transition mineral" has allowed the industry to rebrand itself as a key ally in any transition toward a post-carbon future.[10] As stated in a CODELCO brochure in 2011, "Today's world has a growing need to ... substitute fossil fuels, ... this has motivated the increased use of clean energies, such as solar energy. The physical properties of copper allow this metal to be used in the whole process of capturing this inexhaustible and clean energy."[11] Providing what Pignarre and Stengers label an "infernal alternative," this copper modernity tells us that either we have massive copper mining (with all its negative socio-environmental effects) or we don't have a sustainable future at all.[12]

CODELCO's changing copper modernities, however, do not explain by themselves a corporate performance over forty years that has been described by an analyst of the industry as "nothing short of extraordinary, from any point of view."[13] Along with this public project, the corporation's success rests on its mastery of another, much less visible, kind of technopolitics: *residualism*, a production logic in which valued minerals are—in material terms—only a side effect of the production and management of colossal amounts of mineral residues, mostly tailings. Residualism is at the center of the astonishing increase in production of contemporary copper mining industries, the mineral backbone of the global extension of Western-style consumer society (and Western-style sustainability). Residualism is at the same time the cause of some of mining's more pressing socio-environmental impacts, especially the pervasive emergence of toxic geosymbioses. This consequential logic, the very heartbeat of the copper mining industry, started with the development of a seemingly obscure technology at the turn of the twentieth century.

The start of the twentieth century found Teniente in a state of almost complete abandonment. Having been mined in an artisanal way since the sixteenth century, by 1880 the rich veins that had been exploited until then were largely depleted.[14] What remained were deposits with much lower concentrations of valued minerals, extremely difficult to exploit in a profitable manner with the available processing technologies.[15] The problems faced by Teniente were echoed in other mines around the country, resulting in the almost complete collapse of the Chilean copper mining industry by the end of the nineteenth century.[16]

This decay was highly paradoxical, as copper at the time was more in demand than ever. The massive process of electrification happening worldwide not only depended on the availability of technologies to produce and consume the electricity; crucially, it also needed a medium through which they could be connected. After various options were tried, copper proved to be the most cost-efficient element to perform this task. On the one hand, its atomic structure, based on a single electron, made it a very good conductor of electric energy. On the other hand, copper was much more abundant than other conductors such as gold and silver. As a consequence, "by the beginning of the twentieth century, copper wire was an essential material, found in almost every device and installation known to the electrical industry."[17] Capitalism, especially electric capitalism, would be unthinkable without copper. Operating as a counterpart of the "light modernity" of materials such as aluminum, copper offered the heavy wires that capitalism needed to extend its global reach.[18]

However, from this very atomic structure derives also one of the key problems regarding copper mining, given "that outer electron also means that copper atoms easily bond to many other atoms so that it is rarely found in pure form. In most copper ore deposits around the world the desired metal is bound up with many other less valuable elements, including iron, sulfur, and arsenic, to name but a few of the most common."[19] Copper is utterly promiscuous, forming chemical assemblages with several other components. This promiscuity is highly problematic for its industrial usage, as its requires the mineral to have a purity of 98 percent and above. So there is usually a great distance between copper as present in geological strata and the copper wiring of contemporary capitalism.

This demand for purity put the copper mining industry in a conundrum at the beginning of the twentieth century, as most mines worldwide were facing similar problems as Chile, with the richest veins of copper rapidly depleting. What was left to be mined was mostly porphyry copper deposits, in which copper is found deeply entangled with several other elements. To make matters worse, copper processing was done through a technique known as gravity separation, consisting simply of grinding the extracted material to the smallest possible particles and then shaking the resulting gangue in order for the heavier part—the minerals—to fall to the bottom and be extracted. This technique was relatively simple to implement but extremely inefficient, resulting in the loss of important amounts of the valuable minerals in the resulting waste. Given these low rates of extraction, in the parlance of the industry, by the end of the century there were multiple mineral deposits but few ores, as the ore is the only "economically exploitable section of mineral deposits."[20]

Multiple actors in the industry were searching intensely for a new way to process minerals, one that would allow them to turn such deposits into ores. By the end of the 1880s, the first patents for an alternative method of mineral separation were presented, all introducing procedures based on adding certain chemical reagents to a liquid solution formed by the crushed ores plus water, a mixture that made the minerals in the solution float. Shortly afterward, the first companies specifically centered on exploiting such patents were formed, prominently the Minerals Separation Ltd. Co., established in London in 1903, and the Zinc Corporation, established in Australia in 1905.

Paradoxically, the first proper application of this new procedure occurred not during an attempt to deal with ores but in its very opposite: mining waste. The more than a decade of active processing using gravity separation at the massive Broken Hill mine in Australia had produced several million tons of tailings.[21] The fact that this discarded material usually contained high concentrations of valuable minerals, especially zinc and silver, encouraged many of those involved in mining at the time to perceive that "millions of pounds worth of ore thus went to waste."[22] Inspired by the early patents, technicians and entrepreneurs at Broken Hill started an intensive search for a workable method "to recover the unrealized profit that lay in the tailings."[23] Then waste, from the traditional view of it as merely valueless discards—something to throw away as fast and cheaply as possible—started to be seen as something utterly different, as a source of further value extraction.

After several failed attempts, a breakthrough occurred in 1904 when personnel from Minerals Separation tested at Broken Hill a method "wherein an even smaller amount of oil was added and the pulp was violently agitated to entrain air because the sulfides were carried out into a froth and removed."[24] The success of this method in recovering a substantial share of valued minerals from the waste led to a patent early in 1905 for a technique in which "a large proportion of the mineral... rises to the surface in the form of a froth or scum which has derived its power of flotation mainly from the inclusion of air-bubbles introduced into the mass by the agitation."[25] Flotation was born.

Initially, the focus on developing a technology for waste processing was so intense that many at Broken Hill "largely failed to recognize the implications of the new process.... that flotation was far more than simply a solution to an immediate problem: it was one of the most significant advances ever made in ore treatment."[26] Only when the successes at Broken Hill were analyzed by technicians of the Zinc Corporation was the utility of the flotation method to extract mineral directly from ores identified.[27] This realization started a frantic and fractious (involving several lawsuits in multiple countries) competition between the two corporations to develop, patent, and commercialize the technique.

In a very brief period of time, even by today's standards, flotation radically reshaped the mining industry worldwide. As summarized by Mouat:

> Flotation expedited the dramatic shift to non-selective mining methods, allowing vast underground areas to be sent to the surface for treatment or, alternatively, mechanized excavation to be carried out in huge open pits. Those with the industry's critical skills were now university-trained engineers and metallurgists.... Flotation's development also illustrates the ways in which technology assisted the emergence of modem business forms, whose success often relied upon applied science, sophisticated engineering, and an elaborate managerial structure.... [As a consequence] flotation hastened a fundamental reorganization of the industry, affecting the nature of work as well as facilitating the rise of big business.... [M]ore and more mining companies became city based corporations with diversified holdings in mining, smelting, and other interests, replacing the earlier corporate model which typically was based on a specific mineral deposit in a particular region.[28]

Flotation made profitable—or turned into ores—hundreds of mineral deposits worldwide that had been deemed to have too small concentrations of valued minerals to be exploited. This was especially critical for copper mining, as flotation made it possible for the very first time to exploit profitably the vast porphyry copper deposits found worldwide. From the former perennial search for new high-grade deposits, interest shifted toward the extensive exploitation of known low-grade deposits, trading purity for quantity.

In this shift, the industry itself was utterly transformed. Up to that point, most mining had been carried out using basic technologies that had changed little for centuries, and based on the informal "professional judgment" of longtime miners with limited formal qualifications.[29] In most cases, each mine was owned by a single company focused mostly on direct mineral extraction. In contrast, the profitable exploitation of ample low-grade deposits required an extended investment in infrastructure and a smaller but highly qualified workforce. Such a large investment could only be afforded by increasingly large corporations, usually adopting a multinational mode of operation and financing. Tellingly, the corporations that were closer to the development of flotation were those that ended up profiting the most from this new arrangement, their patents allowing them to become global brokers of the technique.[30]

Flotation—a technique initially developed to treat mining waste, not ores—was the crucial development that allowed the emergence of contemporary mining as we know it. The copper in most geological deposits worldwide was impure, too promiscuous to be exploited profitably. It only became ore once flotation was made extensively available, allowing the massive treatment of previously unexploited low-grade deposits, increasing tenfold the global availability of cheap minerals. As Mouat recognizes, "It is not overstating the case to claim that flotation's development was of central importance to the smooth functioning of the global economy, for without it metals such as copper, lead, and zinc would have become increasingly difficult to produce and their price would have risen as a consequence."[31] Following Mumford's well-known dictum of mining as more "bound up" with the development of modern capitalism than any other industry, flotation could well be seen as one of the most influential technical innovations of the twentieth century, although it is rarely recognized as such.[32]

BRADEN'S GAMBLE

The July 13, 1905, edition of Chile's *Diario Oficial* included a brief notice stating that "exclusive privileges are granted for nine years to the Minerals Separation Limited corporation to use in the country an improvement in the concentration of minerals through humid ways of its invention."[33] Coming only a few months after the developments at Broken Hill, in a time of still fragmentary global interconnection, this early Chilean patent shows the degree to which flotation took the mining world by storm.[34] This rapid diffusion was also due to the emergence of novel, aggressive, and (increasingly) transnational mining corporations such as Minerals Separation, which not only rapidly patented flotation in Chile, but also already had an exclusive representative in the country with the aim of both showing a prototype to interested parties and, as described in a publication of the time, "defending the rights of the patents like a cornered cat [*gato de espaldas*]."[35]

These developments had not passed unnoticed by actors in the local industry, among them an Italian engineer called Marco Chiapponi. A colorful character, Chiapponi had been based in Chile since the 1880s, mostly working as a technical consultant in several mining operations. In the late 1890s he was put in charge of selling several inactive mines in the central Andes, notable among them the almost abandoned Teniente mine. His efforts were unsuccessful until November 1903, when he sent several letters to the American engineer William Braden—whom he had befriended at a mining fair a few years earlier—explaining in detail about the mineral deposits at Teniente and their potential.[36] Braden was enticed enough by the letters to travel to Chile later that year, visiting Teniente with Chiapponi. Being familiar with the latest technical developments in the industry, Braden recognized the huge potential of this vast low-grade deposit but knew it would involve the construction of costly infrastructure. For this reason, he returned to the United States to look for financial backers for the enterprise, after which the Braden Copper Company was created at the end of 1904.[37]

In April 1905 the first basic work started at Teniente, mostly centered on building communication and transport infrastructure, a challenging task given the location of the mine 2,700 meters above sea level in the Andes, more than 40 kilometers away from any proper road. At the beginning, the extracted ore was processed using the traditional gravitational method. However, its low grade meant that in its first years the mine "could not send enough copper to the United States to continue its operations."[38] This situation made the financial standing of Braden Copper quite precarious, even threatening bankruptcy if production were not increased substantially in a very short period. Besides looking for extra funding, Braden decided to take a gamble to save the mine: use flotation.[39]

Despite a frenzy of speculation, flotation at the time had not been introduced in any large-scale porphyry copper mines worldwide. Most of the applications of the process were highly experimental and case specific, revealing that in fact so "little was actually known about it that it was [not] easy to devise a theory that could not be convincingly disproved."[40] Flotation's

FIGURE 5. Teniente's flotation plants, 1936

main industrial successes had been in treating zinc mining tailings, so to use flotation at Teniente would be to enter uncharted territory. In cooperation with Chiapponi, in 1908 Braden contacted Minerals Separation in London to explore the possibility of using its flotation device to process Teniente's ores.[41] After some positive interchanges, in 1911 he sent a sample of 25 tons of Teniente ore to be processed at the Minerals Separation trial plant in London, an experiment that produced highly encouraging results, proving "conclusively that the ore was amenable to concentration."[42] After personally witnessing the successful treatment of the ores, Braden wrote to the mine owners "that flotation would prove the solution of the problem of efficient concentration," and with their backing, he began discussions with Minerals Separation executives about the construction of a plant in Teniente.[43] Minerals Separation, for its part, was very interested in testing its device in a copper mine and so offered to cover all the initial costs of the installation; a 400-ton flotation device was shipped from New York in early 1912.

The prototype was first tested in Santiago and, after it provided the expected results, it was finally moved to Teniente in mid-1912.[44] For most of those involved, the experiment proved to be a brilliant success. According to calculations made by Folchi, after the installation of flotation processing, the mine's production substantially increased, from less than 5,000 tons of refined copper in 1912 to more than 25,000 tons in 1918.[45] This represented 30 percent of the total copper production in Chile and more than 3 percent of global production.[46] For this reason, not only were Minerals Separation's flotation technologies adopted, but they swiftly replaced gravitational concentration for the whole mine, remaining operational until 1956, as can be seen in figure 5.

This increase provided the urgently needed financial lifeline for the struggling Braden Copper Co., which now could be sold to the US-based Kennecott Mining corporation for a substantial profit. Besides the local impact, on the international level "most copper companies wanted to [use flotation]... after its worth had been proved at Braden Copper."[47] This extensive spread of flotation not only increased the capacity to process low-grade ores and massively increased copper production worldwide; it was also centrally related to the emergence of a new kind of mining corporation, capital-intensive "business enterprises" increasingly based on a mixture of technological sophistication and transnational financialization, progressively interconnecting single mines to massive "planetary mines."[48]

TAILINGS ARE BORN

Besides allowing a massive increase in Teniente's copper production, the introduction of flotation had another key effect. As seen in figure 6, flotation basically is a technology for sorting two kinds of materials. Emerging on top of the device thanks to their association with air bubbles are minerals of several kinds, especially copper particles in the case of Teniente. These minerals are then extracted or, as it is usually referred to, "harvested" to be sent for further processing, their ultimate sale being the main reason for the whole enterprise. These minerals, however, constitute only a tiny fraction of the materials produced through this process.

The second key effect of the introduction of flotation at Teniente in 1912 was the production of a novel chemical compound: tailings. Although mining has produced different kinds of waste from its very beginning, the compound that is currently known as "tailings" is inseparable from the flotation process, being a total novelty when they began to be pumped out of the flotation plant in 1912.

In order for flotation to extract the larger amount of the valued mineral, the extracted ores must be intensely crushed and ground, resulting in tailings that are quite unstructured, similar to very light, sandy soils. This loose structure resulted in a compound that is quite geochemically unstable. At the micro level, this instability manifests in the ease with which tailings can produce different unwanted chemical reactions, especially acid mine drainage. At the macro level, it is manifested in the recurrence of tailings dams' being affected by the still poorly understood phenomenon of liquefaction, in which under

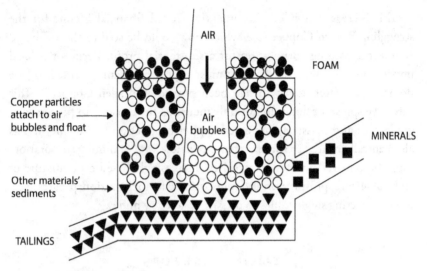

AIR

FOAM

Copper particles
attach to air
bubbles and float →

Air
bubbles

MINERALS

Other materials'
sediments →

TAILINGS

FIGURE 6. Diagram of the flotation process

certain external stress (like an earthquake), accumulated tailings suddenly lose their stiffness, rapidly spreading outside their dams.[49] In doing so, they become particularly slippery, "escaping modern production and regulatory systems."[50] Tailings are extremely leaky, always challenging containment.

The introduction of flotation at Teniente allowed the development of mineral processing "of a colossal scale in comparison with was normally seen in Chile."[51] This scale, not surprisingly, produced massive quantities of tailings, novel compounds that no one at the time really knew how to treat safely. As a consequence, starting in 1912, decades of mineral processing at Teniente were marked by a continuous series of environmental disasters involving the mine's tailings, especially resulting from the sudden collapse of tailings dams, usually leaving a trail of destruction and death on their way downstream.

A main cause for these repeated disasters was that, up to the 1950s, tailings dams were designed and built in the simplest and cheapest way possible, usually assuming the form presented in figure 7. Set on a small foundation made of solid material, usually rocks, at the base of the structure (*terraplen primario*), the rest of the dam wall was made up of tailings, whose progressive dry-up was seen as sufficiently solid to act as the wall holding the rest of the material. This situation only started to change in 1965, after the collapse of a tailings dam in a mine north of Santiago killed more than three hundred people. The subsequent public uproar made tailings a matter of public

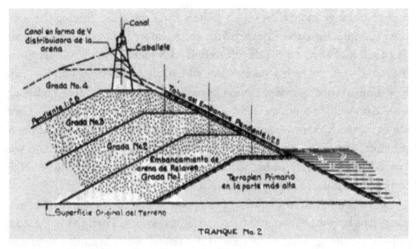

FIGURE 7. Sketch of an early tailings pond at Teniente, 1935

concern in Chile for the first time, leading to the introduction of the first nationwide safety regulations regarding tailings dam construction.

As a result of the repeated disasters and growing public and regulatory pressure, Teniente was forced to progressively change its operational focus. To the traditional (almost) exclusive focus on the extraction, refining, and circulation of valued minerals, it had to add a growing emphasis on waste management, as the ever-growing masses of unstable tailings presented the company with complex technical and environmental challenges. This increased attention paid to tailings was related not only to safety issues, but also to the possibility of reprocessing the tailings, as they always contain a certain amount of valuable minerals that can be extracted by novel means.[52] This emphasis on waste management acquired an especial intensity in the late 1970s, when it was clear that the current dam had only a few years of operational life left. If the residual logic of extraction materialized by flotation/tailings was to have any projection into the mine's future, a new place for a massive dam had to be found, and fast.

MAKING NONLIFE AT CARÉN

The Carén Creek basin rapidly emerged as the preferred option among those being considered for placing Teniente's new dam. Despite being located in a mountain range more than 90 kilometers from the mine, forcing the company to build a massive infrastructure to transport tailings there, the basin had a series of advantages difficult to find elsewhere. First, the geography of the basin—formed by high mountains with a relatively narrow entry—made it possible to contain massive amounts of tailings by only building a relatively small dam wall. Second, the total surface to be occupied by the dam—2,000 out of 22,000 hectares—was just a small fraction of the total size of the basin, substantially minimizing its environmental impact.[53] Third, the basin was located in its entirety inside the massive Hacienda Loncha, whose single owner was willing to sell to CODELCO.[54]

Last, but not least, the land's dispossession from its original human inhabitants had begun well before the corporation arrived. First, the indigenous population living there at the time of the arrival of the Spanish conquerors had been expelled from the area in the eighteenth century.[55] The subsequent introduction of the hacienda system, in which peasants working the land had no ownership of it, extended that dispossession over time. Thus a great deal of the violence related to dispossession would not fall on CODELCO's shoulders.

The dispossession, however, was not complete. At the time of CODELCO's arrival the basin contained not only mountainous areas, but also ample agricultural and pasture lands, on which around five hundred people lived and worked. Although they had no property rights over the land, most of them had been living there for generations, developing strong historical and emotional attachments. As can be seen in figure 8, Loncha was not only their place of work or residence.[56] It was a place of traditions and communal celebrations, some involving hundreds of people. It was a place to fall in love and raise a family, a place to enjoy some leisure time and mourn the dead. It was a place of belonging and identity for most of its inhabitants.

Given these attachments, the inhabitants' sudden removal from the area—from what had made up their lives until then—was not taken lightly, as remembered by Fabiola Varas, one of the former inhabitants of the hacienda:

When ... [the owner] sold the hacienda to CODELCO, they started to demolish all this area, the trees, the houses, they displaced all the people, I mean, ... [the owner] gave a little money for the people to buy something on other sectors, and most the people left ... but some people remained, their animals remained, and they started to turn down their houses with machines and the people had to run, leave in a haste, because they no longer had a place to stay, their animals remained there, thrown there, dogs and things like that.

FIGURE 8. Snapshots of everyday life in Hacienda Loncha

Before starting to receive several thousand tons of tailings per day, a significant part of the Carén Creek basin had to be rapidly turned into a space of nonlife. This process meant utterly obliterating the landscape of the area, starting with the disappearance of Carén Creek itself. It also meant laying waste to all the human, animal, vegetal, and arboreal lives existing around it, violently displacing bodies and tearing down their surrounding materiality. They were treated as waste, as a valueless remainder of the past incarnation of the area that must be hastily removed and forgotten. Humans, animals, crops, memories, trees, traditions, houses, all must go. Everything was done with the final aim of enacting a perfect emptiness, a space of nonlife that could reanimated by the soon-to-be-coming tailings.

The violent creation of this space of nonlife appeared to have paid off on that day in November 1986 when tailings finally reached Carén. At last, CODELCO's ambitious copper modernities—especially its current "sustainable" version—were matched by a waste depository on which the residualist program could fully unfold. Teniente now had a massive logistic network to mobilize and collect the value stored in refined copper, becoming an honorable member of late capitalism's planetary mine. The vastness of the dam—with estimates projecting a repository that could last more than one hundred years—seemed to confirm the continuity of the residualist logic that had guided the mine's operation for the past seventy years, projecting it well into the future. Carén could be seen as the conclusion of the project started in 1908, when Braden contacted Minerals Separation in London, and Teniente became the first porphyry copper mine worldwide to use flotation on a grand scale.

At Carén tailings could be laid to rest seemingly forever; inert, frozen, embodying the ultimate nonliving space of late capital extraction. They would also become largely invisible except for a few technical personnel, especially disappearing as a matter of concern for society at large. As a consequence, an industry that mostly produces residues can be instead seen as one solely producing variable amounts of valued minerals. In doing so, residualism emerges as the perfect mining technopolitics, one of the more comprehensive ways to enact spaces for nonlife in resource extraction, the opaque (but utterly necessary) counterpart of the global logistics circuits of value and desire.

Sadly, in practice residualism does not work as claimed. Carén was never the perfect nonliving place, not even at the very beginning. As we will see in the rest of the book, even before the tailings reached Carén, multiple living entities—some displaced during construction, several others brand new—became entangled with them. Such geosymbioses made it impossible for tailings to simply rest forever; instead they embarked on a nonstop process of forming unlikely alliances with multiple living organisms, from members of local communities to algae. Tailings' geosymbioses do not mean that residualism disappeared or collapsed, though. Materializing the well-known resilience of capitalism, at Carén the residual logic rapidly adapted and evolved, trying to bring the geosymbioses into the project of nonliving stillness. In doing so, residualism became a weird kind of animism, formally recognizing the existence of geosymbiotic entanglements but in practice always on the lookout for new ways of governing them *as if* they were nonliving.

TWO

Carp, Algae, Dragon

SEDIMENTATION

In contrast to the efforts necessary to start tailings flowing into Carén—from building Chile's longest tunnel to violently ending local life stories—everyday life at the dam seemed marked by radical stillness. At least this is what we witnessed upon first arriving there in 2014. After the constant activity needed to move tailings through the canal, the dam itself seemed like the epitome of tranquility.[1] Standing on the wall we saw a mass of light-green water, its stillness only disturbed by an infrequent gust of wind rippling its surface. Nothing seemed to be happening at the dam, no matter how much time we spent looking at it.

This impression was reinforced by a common discourse among the directives of the waste management area of Teniente stating that managing tailings inside Carén was relatively straightforward. As Raúl Cuevas, one of the higher directors of the area, recalled to us one afternoon in June 2015 in his office in Teniente headquarters, in this process "we use the laws of nature, nothing more." In Cuevas's telling, as soon as tailings enter the dam "a sedimentation of the solid fraction of tailings occurs.... [T]he speed of movement inside the dam is too slow to carry the solids, so a decantation occurs that is called the sedimentation process." The selection of this term was not arbitrary. In geological literature, *sedimentation* refers to the process "when particles in suspension settle out of the fluid in which they are entrained and come to rest against a barrier, which is typically the basement of a waterway."[2] Sedimentation constitutes one of the most important geological processes because it is responsible for the formation of a wide range of geological strata, especially via river sedimentation.

In mining waste management, sedimentation is also capital, as it is the central expected effect of infrastructure such as a tailings dam. Given this fact, it is not unexpected that a review of technical literature on the matter affirms simply that "dams are best thought of as purpose-built sedimentation lagoons."[3] Dams are made for tailings to sediment, letting go of most of their water content and becoming solids, ready to start the geomorphological processes that turn them, in deep time, into just another underground stratum.

Adopting a Deleuze/Guattarian frame, sedimentation can be seen as a massive geological "territorialization" force, or a process through which a mass of heterogeneous and loosely connected materials called tailings come to be seen as progressively forming a coherent geological whole in the form of sedimentary rock, a kind of rock covering more than 70 percent of the earth's land surface.[4] Sedimentation creates an impression of the dam as a (future) stable whole, marked only by slow-paced geological processes and well-defined objects. By referring to what happened inside Carén as a sedimentation process, Cuevas was connecting the dam and its tailings with an extensive array of different, but utterly "natural," geological phenomena. "This is a natural process," he claimed "in which only the particle's gravity and water particularities participate, and then a separation occurs . . . between the solid and the liquid components [of tailings], and this is manifested in the sedimentation."

Following a view of geology as a political practice, the idiom of sedimentation at Carén clearly appears as a way to naturalize the extractive process.[5] This naturalization is quickly followed by normalization, by a view of what happens at Carén as utterly normal and hence uncontroversial. Beyond being merely a descriptive device, Cuevas's "natural sedimentation" of tailings takes on the strategic role of complementing the processes seen in the previous chapter, helping CODELCO to turn residualism into something natural. What was ungrounded by force, for profit, is now safely grounded again by nature, closing the geological circle.

This naturalization masks the radical violence of residualism, regarding the earth itself and its inhabitants. At Carén, sedimentation conceals all the things that were crushed under tailings, from aquifers to memories of past inhabitants, and legitimizes the formation of this grey flatness of desolation, where geological "natural" processes will heal the wounds of extraction in deep time, a time so deep that no one really cares about it now; it is not their business. Given their apparent inertness, such capitalistic "deserts" are places of structural forgetfulness, where any kind of future environmental impact, of fast or slow violence, and its derived responsibilities, is covered in a thick layer of geological inevitability.[6] As a result of this operation, Carén becomes "a material enactment of forgetting", or a huge piece of infrastructure for non-seeing, for learning to ignore the geosocial consequences of residualism.[7]

In acting this way—adapting freely from land artist Robert Smithson—the sedimentation of the *mine* easily turns into the sedimentation of the *mind* for personnel such as Cuevas.[8] Falling under the spell of this second kind of sedimentation, tailings' "movement seems motionless, yet it crushes the landscape of logic under glacial reveries. . . . The entire body is pulled into the cerebral sediment, where particles and fragments make themselves known as solid consciousness."[9] As a result of this double sedimentation—both material and perceptual—any possibility of questioning residualism, of facing its utter violence, is crushed under the weight of "natural" geological processes.

From the standpoint of the dam's wall, with its awe-inspiring vistas, it was easy for us to fall under the spell of this sedimentation of the mind. We just stood there, like seekers of ruin porn rejoicing in the strange beauty of residualist destruction. But as soon as we left the lookout, things started to change, because lurking above, below, and even within the tailings, we found multiple challenges to sedimentation in the form of geosymbiotic entanglements, from the molecular to the geological, that openly contradicted the image of the dam as only marked by geological sedimentation. It all started with *Cyprinus carpio*, also known as the common carp.

THOSE SHITS

One afternoon in December 2014, we visited the dam with Leonardo Carvallo, a longtime employee of the waste management area of Teniente. We drove up along the dam's wall until we reached a lookout, from which the whole expanse of the deposited tailings could be seen. We went to this location so that he could explain to us the functioning of the discharge tower, a structure that extracts the water resulting from tailings' sedimentation and sends it to a treatment plant. We observed the tower, a thin concrete structure emerging from the water right next to the dam's wall, for a few moments, but before he could start the explanation, something else caught our attention.

Pointing at the surface of the water around the tower, Carvallo told us *"those little spots that you see there are fish."* At first, it was difficult to see the spots he mentioned, since we were several dozen meters above the water level

FIGURE 9. Carp at Carén Dam

and the reflection of the sun on its surface made it difficult to see anything. But after becoming accustomed to the glare, we started seeing the "spots," little gray points on the water's surface. On our way to the lookout, Carvallo had mentioned that there were carp in the dam's lagoon, but we did not expect to see so many of them, dozens of grey spots moving around the tower's base. He seemed quite eager to show them to us, so we drove closer to the tower to get a better look at them. Once there, we could see that the carp were quite large, averaging at least 80 centimeters in length, and ubiquitous, especially at the foot of the tower, where dozens and dozens of them could be seen happily pullulating below the surface (see figure 9).

After watching them in silence for a while, we asked Carvallo if he knew where the carp came from. He answered, in his usual straightforward style, "I don't know where those shits came from." The arrival of the carp at the dam lagoon was a total mystery for everyone at CODELCO, given that it was (seemingly) completely closed off from its environment, its only source of water besides the tailings being a little stream coming from the surrounding mountains, which ran dry most of the year. Furthermore, the high concentration of several potentially toxic minerals in the lagoon's water made it highly unlikely that any kind of fish (or any kind of animal) could survive there.

But carp are highly resilient beings, able to resist a wide array of toxins, especially those generated by human activity. Indeed, "there is no freshwater fish stronger and more resilient to human management than carp."[10] As a result, carp have become one of the most extensively invasive fish species worldwide, significantly challenging most kinds of native fish populations in the process. As vivid testimony to their resilience, at Carén carp not only had arrived by unknown means and survived in normally toxic conditions, they had also thrived in them, becoming a large community.

However, this resilience and pervasiveness do not mean that carp are appreciated. On the contrary, most of the time the carp is thought of as "the quintessential trash fish, an exotic, invasive bottom-feeder that thrives in polluted environments."[11] Carp are usually unwanted, actively despised, inedible, and ugly, a pest that must be controlled and (hopefully) eliminated, a lively reminder of the huge damage caused to global watersheds by human activity. Normally our only way of engaging with carp is trying to (usually unsuccessfully) exterminate them. Carp, it can be concluded, are the ultimate unwanted "companion species" of humanity's freshwater messes.[12]

Following this latter trend, Carvallo's disparagement of carp as "shits" makes his engagement with them fairly limited, as he expressed that day.

FLORES: Have you ever fished for one?

CARVALLO: Yes!

FLORES: Really?

CARVALLO: Yes. From here they look gray, dark, but when you catch one they look golden, very golden and light.

FLORES: They are big...

CARVALLO: Yes.

FLORES: And what do you do with them? Do you eat them?

CARVALLO: No! We take them, we see them, all that shit, and we leave them in the water.... Look how they are here [points to a lot of carp gathered close to us].... From here you can get a good picture!

Besides sometimes catching an exemplar just to pass the time, Carvallo and his colleagues avoid any further contact with the carp; he clarifies immediately that he does not eat "those shits." Carp inside the dam are despised, utterly useless, and only tolerated because they are deemed harmless. Although they have spread out over the whole lagoon, they do not seem to interfere either with the sedimentation of tailings or the treatment of water, so they have been allowed to remain there, becoming little more than a weird curiosity to entertain visitors, as Carvallo suggested when he invited us to take a good picture. Other uninvited guests to the dam did not encounter such tolerance, however.

One afternoon in March 2015 we were with Fernando Espejo, a process analyst at the waste management unit, in his office at Teniente headquarters talking about the treatment carried out on the water released from the Carén dam. While speaking about the seasonal change at the dam, he commented that the highlands above it present an important amount of biodiversity, even housing a public natural reserve. Given this, is it not strange that during the rainy season, "several organisms come down from the hills, organic [elements] and living things which stay in the dam." A central position among such entities is occupied by algae—probably of the green algae kind, as they are the most common freshwater algae in Chile—microorganisms that form the very foundation of most aquatic ecologies by being a primary source of organic matter.[13] Upon arrival, their presence in the lagoon is not even noticed by CODELCO personnel. But the situation changes radically when the spring heat arrives and "the precise, exact conditions for the growth of some microorganisms occur, especially algae."

This process, repeated year after year, has led to the recognition of two cycles within the dam, as described by Cuevas: "I do not know if it is because algae become romantic or what, one [cycle] happens in the autumn [April–May] and the other in September, in spring." During these cycles, usually lasting one to two weeks, not only can algae be seen floating everywhere in the dam; more importantly, the very characteristics of the water change. As he continued, normally they have "an utterly pure water, besides the molybdenum, then the water goes into the [processing] plant and all its equipment functions as if it were a laboratory. But when this blooming appears, . . . it makes [the] water dirty, between brackets, for [their] ends, and as [it is] . . . of natural origin, this organic issue also alters . . . [the processing] plant; specifically, the [molybdenum's] extraction rate decreases."

Besides deep-time sedimentation, the dam is activated by the continual extraction of water from it, water that goes on to form Carén Creek. This extraction responds mostly to the need to limit the dam's growth rate, so it can remain operational until its planned closure date of 2064. Before being released, though, the water needs to be treated, mainly by removing certain potentially toxic components, especially molybdenum. As part of the compensation package that CODELCO was forced to administer after the 2006 spill (discussed in chapter 5), the corporation built a molybdenum treatment plant, known internally as PAMo, where the water is processed in order to adjust its levels of molybdenum and pH according to the maximum concentrations allowed by Chilean environmental regulations. Designed and managed by the French company Suez Degremont, PAMo has been promoted from the beginning as an example of CODELCO's technical excellence and environmental commitments. Given this, anything that conflicts with the plant's "faultless operation"—as described in an internal brochure in 2011—is treated with utmost seriousness, something to be dealt with as quickly as possible.

Here enter the *romantic* algae. Twice a year, especially in the spring, the dam experiences massive algal blooms. According to the literature, such blooms are caused by an imbalance in "physical, chemical and biotic factors . . . [producing] an excess of phytoplankton accumulation . . . that lead to visible discoloration of the water."[14] The exact cause of the blooms at Carén—especially the one in autumn—had been impossible to determine; "even specialists [have tried to understand them without success]. It is very difficult to define, to understand," Cuevas commented. These blooms usually cover large extensions of surface water in and around the dam, as can be seen in figure 10, taken at one of the dam's discharge canals.

When algae become "romantic" in such a massive way, the PAMo cannot continue working "as if it were a laboratory," following Cuevas. Contrary to expectations, industrial wastewater at Carén becomes polluted not by potentially toxic compounds but by the presence of organic entities such as algae; they are the "matter out of place" at the dam, importantly affecting PAMo's "faultless operation."[15] Luckily for Cuevas, the solution for this conundrum was at hand: "I have to sacrifice the plant's efficiency and put more equipment to work, as simple as this."

However, for the people working at PAMo, the solution was not that simple. As Espejo recalled during our conversation, algae not only pollute the water coming from the dam but even "inhibit the action of some reagents. . . . It is a purely chemical issue. [In normal times] you add reagents to water and have an impeccable functioning, with everything under control, but then [during blooms] . . . it does not work as expected, and it is due to these organisms." During blooms algae entangle with some of the reagents used at PAMo,

FIGURE 10. Algal bloom in Carén Dam, September 2015

causing the latter to stop capturing molybdenum particles from the water. This entanglement was completely unexpected, uniting algae whose ancestry in the Carén basin probably can be traced back for centuries with reagents coming from the laboratories of one of the world's leading wastewater treatment corporations.

This geosymbiosis starts a frantic period of looking for a solution, with little success: "We look [at] the pH, we look at the reagents, this one yes, this other no.... [But] over two weeks... we are fucked, and we have no idea what's happening." Algae-reagent entanglements occur according to their own principles, showing an utter indifference to Teniente's plans. This indifference ensures that, during the blooms, there is little control over the quality of the water released from the PAMo. As Espejo indicates, "We are releasing water with 0.5 [ppm of molybdenum per liter], and suddenly we have 0.8, almost 1! Fuck! But we are adding the same [dose of reagents], we have the same pH, ... we have everything in the same way!" Algal blooms all but invalidate the processes enacted to meet CODELCO's environmental regulation obligations, making likely the emergence of further "out of place" substances, such as water with high concentrations of molybdenum in Carén Creek or, even worse, contaminated animals or humans downstream.

Carp and algae are utterly despised entities, "*those shits.*" Such dislike is not related principally to the fact that they are deemed worthless and ugly, an unnecessary and invasive species. Beyond this, they are despised (even feared) because "the failure to manage... these animals with the best scientific research and control methods reveals our all-too-human limits and makes us confront the fact that much of the natural world remains a mystery to us."[16] Carp that are able to thrive in highly polluted water and algae that interact with reagents present Carén personnel with geosymbioses that refuse to align with the sophisticated control programs that supposedly characterize modern mining.

To CODELCO's advantage, these entanglements' defiance of its control was limited by space (carp remaining inside the lagoon) and time (algal blooms emerging only twice a year for a few days). For the remainder of the year, they could still assume that there was "faultless operation," with the dam working as a massive piece of infrastructure for the sedimentation of extraction, the geological component of residualism. This satisfactory outcome was much more difficult to achieve with another vital entanglement we found at Carén: the dragon.

THE DRAGON

"The dam, for me, is a dragon" said Cuevas to us out of the blue, "it is there, quiet, and you should leave it quiet, nothing more, ... so that it doesn't wake up, it is the same as in The Hobbit movie, the same theory, you can do a lot of things, but you must not wake him up." This sudden appearance of a mythical creature was surprising to us, especially given who was making these comments. Cuevas was a civil engineer who had arrived at the waste management unit of Teniente in the wake of the 2006 spill. As a key part of an extensive modernizing program carried out by CODELCO since the early 2000s, his main task was to completely transform the unit in line with the most up-to-date waste management technologies available. Cuevas's reform program could be seen as an attempt to turn Carén into a proper inert "desert" after the spill and the social unrest it caused.[17]

The dragon could not be further from such a program. Traditionally, dragons are monsters that "appear . . . as embodiments of death and damnation, and their appearance is calculated to terrify."[18] Contrary to the modernizing notions championed by Cuevas, the dragon can be seen as a direct heir of traditional notions of the monstrous emerging in mining areas in Latin America, as noted by authors such as Taussig and Nash, frightening figures associated with "images, memories or recollections directly linked with destruction, violence or violation."[19] Through its enactment as a dragon, tailings at Carén were seen "as alive, resplendent with movement, color, and sound."[20] Tailings-as-dragon possessed a certain inner capacity, an untamed quality that clearly signaled the limits of rationalizing programs such as the one headed by Cuevas. As even he recognized, once "the dragon is fully awake, you have no options other than entrusting yourself to the saints, nothing more."

This dragon was enormous and growing every day, due to the never-ending flow of tailings. Besides its sheer size, this monster was also not inert. Like Smaug—the dragon in J. R. R. Tolkien's novel The Hobbit—the fact that the dragon at Carén was sleeping did not mean that it was immobile. On the contrary, the dragon was constantly acomodándose (adjusting itself) in unexpected ways. As Cuevas continued to describe, "every day we put tailings in Carén . . . every day 150,000 metric tons arrive. . . . Then, there is a movement in the tailings, an acomodación [adjustment], despite how big it is. It adjusts, it adjusts." Complying with the logic of sedimentation, the dragon at Carén was constantly acomodándose, permanently adjusting its ever-growing volume in the dam. However, and contrary to the glacial pace associated with geological processes, such adjustments were usually unpredictable. The level of tailings at the depository "should increase 20 centimeters per month, but sometimes it does not grow, it does not grow, it does not grow. It could remain a year and months without moving . . . and then, suddenly, it adjusts."

Instead of following regular patterns, adjustments tended to happen quite rapidly, suddenly mobilizing thousands of tons of tailings toward the dam wall and, especially critical, toward the discharge tower—through which the water resulting from the tailings' sedimentation is sent to the treatment plant—in a matter of minutes.

Through the figure of the dragon, Cuevas was recognizing that large and complex infrastructures such as tailings dams "have dynamics of their own, and this often includes absorbing pressures or changes of input up to a point, then lurching, quickly and unstoppably, into a new state."[21] Beyond expectations of tranquil sedimentation, when seen as a dragon, this process is endowed with dynamics that are impossible to foretell, even with the most sophisticated equipment.

For this reason, the relationship with such a monster is not one of perfect human control, but of careful treatment, starting with taking the "beast's" own rhythms into consideration. As Cuevas claimed, "after a [heavy] rain sometimes the dam rises very little . . . but also, sometimes, from one day to another, without rain, in summer, it rises 20 centimeters. It means that the dragon [carried out] some adjustment, and if it has started to adjust it means that it will keep adjusting for a while." Thus, the starting point in the relationship with the dragon is to respect its mobility, to wait until it stops adjusting. "A stretch from a beast of this size does not last a short time, it could take days. . . . It could last a week and then it stabilizes, it adjusts, and then we restart," usually having to deal rather urgently with the sudden increase in the level of sedimented tailings around the discharge tower. This is a dragon with no small degree of autonomy, a monster that, although it is sleeping, is regularly adjusting, responding to its own continual growth, an adjustment that cannot be predicted or postponed, only endured. Carén-as-dragon is clearly an open challenge to residualism and its aims of total control through deep-time sedimentation; it is a monstrosity whose "very existence [acts as] . . . a rebuke to boundary and enclosure."[22]

However, this respect for the beast's adjustments rests on Cuevas' ultimate belief that they are going to be circumscribed within certain predetermined limits, so "there is no risk that the tailings will go out." The dragon is free to move, but only as long as it remains captive inside the dam. And this is secured through the regular application of controlling measures, especially moving upward the entrance point to the water discharge tower, so the sudden rise in the tailings' levels cannot flow through it. This process, as he concluded, "is not that much work, it is programmable, you can attend to other demands in between." In this attitude we can see Cuevas' technocratic waste management program at play, the belief that the new, expertly devised control systems installed in Carén would ultimately be able to resist the ferocity of the beast in any circumstances, making it impossible for it to escape and unleash its fury on the surrounding areas. Adjustments, from this perspective, are just a temporary nuisance, random accommodations of the dragon before it goes to sleep forever via sedimentation.

In the end, the dragon is still a figure of governance, a kind of geontology through which actors such as Cuevas, in spite of approaching tailings as animated entities with disruptive potential, render them controllable.[23] The dragon does move from time to time, adjusting its ever-growing body, but these movements are always contained within the limits laid out by Teniente's new waste management system. Particularly in comparison with the period before the 2006 spill, the present life of the dragon in Carén appears to be relatively harmonious, in which both the beast and an up-to-date waste management system can coexist in peaceful proximity. In Cuevas' version of events, the dragon is an incredibly powerful beast but is sleeping most of the time. It is a temporal nuisance to residualism, for sure, but if everything is done right the dragon will peacefully sediment into geological strata sooner rather than later.

In a similar fashion to the case of algae, this peaceful coexistence between the dragon and residualism looked quite different at Carén. We were able to witness this one morning in January 2015, when, standing in Carén's control room, we participated in the following exchange between Leonardo Carvallo, Fernando Hermosilla, and Sergio Inostroza, CODELCO personnel who had been working at the dam for a long time, regarding the recent growth of the tailings level at the lagoon and the ensuing need to carry out a new *sellado* on the discharge tower.

HERMOSILLA: The lagoon grew 60 centimeters last month. I mean, it grew 20 centimeters in 10 days and we, with two tilings [*losetas*], are going to be only 18 centimeters above [the 2-meter buffer]. I mean, it could be said that in 10 days more we are going to be below the 2-meter [buffer] . . . with the last blockage we had a buffer of 2.88 meters . . . and now we are going to have 2.18. . . .

INOSTROZA: So, we should make a 75-[centimeter] blockage?

HERMOSILLA: Mmm . . . last time it was one meter. . . . [T]he problem is what happens if the [water] pressure is low afterwards. . . .

CARVALLO: If you do this you run the risk of having to take off the last tiling, and if the thing is glued on with concrete, how are you going to remove it?

HERMOSILLA: Yes, that is what I meant. . . .

INOSTROZA: A pure mess. . .

[Hermosilla keeps making calculations to himself for a couple of minutes, mumbling.]

HERMOSILLA: No, we are fucked. . . .

FLORES: Why do you say that we are fucked?

HERMOSILLA: Because we have too little water, the height of the tower runs out. If we had been 2 meters below, 3 meters below, we could have done a 1-meter blockage, a 50-centimeter, but now we cannot do it.

CARVALLO: We are . . . more than fucked.

As noted previously, water is constantly being extracted from the dam through a discharge tower. In order to avoid tailings reaching the discharge point in the tower, a series of security thresholds has been introduced, specifically a buffer of 2 meters between that point and the surface of sedimented tailings (see figure 11). This distance is continually checked, and as soon as it falls below 2 meters, or is close to it, a new entry point in the discharge tower is opened a couple of meters above and the existent one is closed with concrete blocks, a process known as *sellado* (blockage). Such *sellados* constitute the main procedure through which the dragon and its destructive capabilities are contained within the dam, allowing the waste management system to continue functioning while dealing with this quite strange and unpredictable "beast."

The image that emerges from the exchange between Carvallo, Hermosilla, and Inostroza is quite different from the one described by Cuevas in his office in Rancagua. Here we see not blockages that were programmed and executed in a timely fashion, nor buffers that were never surpassed to keep the dragon sleeping. Here there is a dragon that continually *pilla* (catches) the personnel of the dam unprepared, as they repeated over and over again that day, rising rapidly from one day to the next and trespassing into the 2-meter buffer. Instead of the perfect planning envisioned by Cuevas, managing tailings at this location was more like a game of hide and seek, in which CODELCO personnel played the part of the ones being pursued and forced to flee every few months, reacting to the dragon's latest movements.

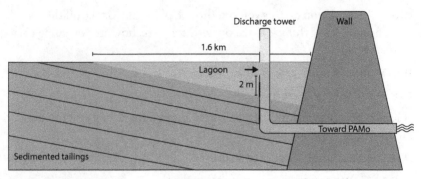

FIGURE 11. Carén Dam's water discharge system

From this position, the dam does not appear to be a sleeping dragon at all. It is still gigantic, monstrous in nature. But it is also fully awake and active, usually in a rather unstable fashion, continually changing its behavior just at the point when personnel thought they already knew how it was going to react next. Instead of the planned and ordered process needed to keep it sleeping proposed by Cuevas, at Carén we witnessed a rather messy and hurried assemblage of heterogeneous components looking to counter the last moves of this cheeky dragon. Cuevas's new waste management program appeared to be openly at odds with such hurried interventions, a divergence that was vividly illustrated by Hermosilla later on in the conversation, when he declared with exasperation: "For me that Mr. Raúl [Cuevas] has not seen the work that is needed. From Rancagua it's so fucking easy to do!"

Hermosilla's rage was not only related to how simple in theoretical terms the implementation of a *sellado* looked from an office in Rancagua, more than 100 kilometers away. For him and several of his fellow workers, the procedure was also a matter of personal well-being. Not only did they work every day under the shadow of the dragon, so to speak; in the case of Carvallo, he actually lived there with his family, in a house near the entrance of the Carén complex. So any unchecked behavior on the part of the dragon, any massive spill, would not only be an environmental disaster but could well end up being a matter of life or death. The dragon was personal, a constant source of anxiety and fear.[24] Certainly things look rather different when you have to face a reckless dragon face to face, when you work (even live) under his shadow and menace, instead of seeing a sleeping beast from a distant office.

Such conflicts became clearer a couple of hours later in a conversation between Fernando Hermosilla and Juan Pinto, another employee, about the lack of concrete blocks needed to carry out the *sellado* of the discharge tower.

PINTO: What the fuck are we doing? Because the [tailings below the] fucking tower are catching up with us, the machine is catching up with us, and then we are thinking "what can we do?" And then [the bosses] say to you that no, that we should stop messing around and delay making the blocks, then we do not make the blocks. . . .

HERMOSILLA: No! We need the blocks; this is why I have been pestering Mr. Cuevas for days!

PINTO: And then I said "you know, Mr. Cuevas, I am going to say something, in one instance, in April of 2006 . . . they cut off the balls of the environmental [area] boss, of the boss of [this] . . . and the boss of [that] . . ." "And why?," he asked me, and I told him "because the fucking tailings from Carén got out! Because the machine caught up with them. . . . And, you know what? This mess happened because of the tower blockage. . . . [W]e need the blocks to seal the tower, or do we leave the crappy thing as it is? Shouldn't we build something?" . . . So for like five years we have been asking for blocks and nothing has happened and it was out of sheer luck that we realized [the level of the tailings] at tower 5 and saved ourselves, remember?

From the perspective of on-the-ground employees, avoiding a spill of tailings outside the dam is a constant struggle, in which the treacherous vitality of the dragon pairs up with bosses who do not understand the urgency of always being prepared to carry out blockages at the last moment, insisting on seeing them as an operation that needs to be well scheduled in advance and (usually) carried out by external contractors.[25] In contrast, ground employees have to deal experientially with the dam, especially with its unexpected adjustments, as a result of which they "have developed their skills at making sense out of the hints contained in the rocks or the streams," a capacity for constant improvisation that is clearly at odds with most of the notions behind up-to-date mining waste management programs.[26]

Instead of experiencing a peaceful, deep-time sedimentation, tailings at Carén appear to be endowed with a capacity to act in strange, dizzying ways, at least from the perspective of the human beings charged with managing them. As a result, the only option is to impose tentative buffers, hoping that they will be able to contain the next adjustment, the only alternative being "entrusting yourself to the saints." Tailings-as-dragons are governed through weakness, a form of control that assigns power to them and weakness to humans but still aims at governing them by establishing a liminal space of precarious control over the figure of a beast that it is necessary to keep asleep.

CARÉN IS ALIVE

After all we had seen there, we were not surprised in the slightest when Fernando Espejo told us in one of our talks that Carén is "extremely alive, there are lot of microorganisms present there, lots of behaviors. . . . [T]he fish that are there, the algae that are there, there is a whole sub-world, subaquatic, submarine, sub-algal, microorganisms, everything, it's a living thing." Talking about Carén as alive was not simply a metaphor, but rather reflected a reality that workers of the waste management unit have to face every day. From algae forming weird alliances with chemical reagents twice a year to a not-so-sleeping dragon always threatening to unravel its fury, the dam was a field crisscrossed by multiple unexpected vitalities: symbiotic and geosymbiotic, micro and macro, seasonal and irregular.

Contrary to its conventional image as the "ultimate sink," a wasteland only verging toward sedimentation, from these vignettes Carén emerges "not as the distant dumping ground of industrial processes of resource extraction and production, but, instead, as a figure for describing hybrid, transformative ecologies."[27] Nothing seems to be actually dead at the dam, not even inorganic compounds such as tailings.

What is more relevant, most of this life is unruly, presenting an ultimate *inhuman* indifference toward human projects. In some cases, such indifference was meaningless or merely amusing to those in charge of the dam, such as the hundreds of carp lurking in the lagoon's water. But in other cases it was much more problematic, even enacting, as in the case of a fully awake dragon, a government through weakness, a form of rule that depends on the environment's indifference to keep existing. In stark contrast with residualist narratives of ultimate human control over nature, such weak forms of power enact "an environmental ethics of vulnerability in which humans are vulnerable to living and nonliving Earth processes."[28]

Returning to the image that opened this chapter, these multiple geosymbioses appear as a force of *desedimentation* of the residualist geologies at play in Carén, and Teniente at large. They undermine the narrative of the dam as a space of nonlife by presenting multiple organic and inorganic ecologies in a location where only "natural" geologic processes should exist. As a consequence, the processes at play in Carén cannot be seen merely as geomorphological, relevant only for technical discussions about geological *deep* time, but as vivid manifestations of emergent and dynamic geosymbioses, occasions for daily responsibilities, (dis)attachments, and dependencies.

Contrary to the tendency of Teniente's personnel to ultimately bemoan such "life," seeing it as a despicable but unavoidable consequence of managing such a large piece of infrastructure, in this book these geosymbioses are taken as the first clues pointing to the development of a more generative politics for extractive operations such as Carén, a politics of unruly aliveness, inescapable entanglement, and human fragility.

THREE

Happy Coexistence

NAMELESS COWS

Matilda is a beautiful cow, to the best of our knowledge. She looks well-fed, with her brown hair shiny and smooth (see figure 12). She is clearly taller than the other three cows on the farm, even though she is the youngest. Upon seeing her that day in November 2014, Tomás Venegas, the farm manager, greets her merrily: "Hello, Matilda!," speaking as if she were a pet. "Listen, you still have food, so don't look at me with such a sad expression!" he says to Matilda. "She is beautiful," he tells us, and we agree with him, to which he emphatically complements: "She is very beautiful!" Venegas tells us that Matilda is his favorite, so he usually gives her an extra bit of alfalfa. Apparently Matilda acknowledges such special treatment, always coming near the fence and mooing loudly when she hears Venegas's truck arriving.

The cow in the next pen does not look as healthy as Matilda at all. It is much thinner and smaller, with messy and tousled hair (see figure 13). It almost does not move during the time we observe it, appearing somewhat shrunken. "This is the one that was in a very, very bad shape," Venegas tells us in a somber tone, "suffering, without being able to move." Presumably the cause of its bad condition was drinking solely the water released into Carén Creek and feeding on pasture also irrigated with that water. Apart from the health problems, the cow does not seem to possess any other remarkable feature, not even having a proper name, only being referred to a couple of times as "La Pobrecita," the poor thing. We continue walking, stopping briefly in front of the next two pens, each one housing a nameless cow. They do not look in such bad shape as La Pobrecita, but they are still clearly slimmer and smaller than Matilda, despite being much older.

FIGURE 12. Matilda

FIGURE 13. La Pobrecita

Matilda and the three nameless cows were the last of a long line of animals, plants, and soils used—and quite frequently abused—in field experiments carried out at Estación Experimental Loncha, an experimental farm created at Carén in 1987 and located 1.5 kilometers downstream from the dam wall. Although the particular organizations, materials, and living beings involved in each wave of research carried out at the station have changed over time, the ultimate objective of the exercise has remained unaltered: a never-ending attempt to validate scientifically what we call the *happy coexistence* thesis.

As already seen, for the dam to continue working as expected until 2064, tailings water needs to be continually extracted from it. After being treated in PAMo, that water is released into Carén Creek, becoming its main—and usually only—source of water. Given this setup, the happy coexistence thesis claims, quite simply, that such water could be used on farms downstream for watering crops and livestock without major disruptions, especially negative environmental or health issues. Or, in symbiotic terms, it claims that any possible geosymbioses emerging from the encounter between tailings water and agricultural beings would not be detrimental for the living beings. If geosymbioses were to emerge, they would be of a commensal or, hope-fully, mutualist kind, resulting in the mutual benefit of all involved. Even the possibility of having parasitic or, god forbid, toxic geosymbioses was outright denied.

As a result, the dam could be transformed from a potentially dangerous disturbance into a source of multiple kinds of conventional agricultural prac-tices, enacting a win-win arrangement between Chile's two main extractive industries, mining and agriculture. This arrangement would produce a situ-ation in which "mining is not the antagonist of the environment, but ... has a very positive impact on it," as stated in a CODELCO publication on the matter.[1] If sedimentation could be seen as the geological program of mining residualism, then happy coexistence should be seen as its biological one.

To validate the happy coexistence thesis, CODELCO needed proof of its existence. Such proofs needed to be strong enough to convince heteroge-neous publics, from a scientific community openly critical about the mining industry's environmental impact to local communities, who watched the establishment of the dam with concern. In addition, at the time national authorities were responding to the growing environmental awareness in the population and the media by establishing the first regulations for the opera-tion of mining waste complexes such as Carén. Given this, the coexistence thesis must not be only a declaration of principles or an educated guess. It should be presented as a *scientific fact*, and the Loncha experimental station would be the main vehicle to produce such facts.

Initiated in the late nineteenth century as an answer to the artificialness of laboratory experiments in the biological sciences, experimental stations such as Loncha looked to produce "a kind of science that blurs the line between lab and field."[2] On the one hand, knowledge production at experimental farms would still be guided by the application of conventional experimental research procedures, lending an important degree of scientific validity to the results. On the other hand, the location of those procedures in situ, that is, where the studied problems originate, experiencing several defining environmental conditions, theoretically would give the collected data a degree of "naturalness" impossible to reproduce in a lab. In clear contrast with the conventional image of the lab as placeless and universal, the value of the experimental farm came "from the ordinariness of its landscape."[3] The combination of rigorous experimental procedures and realistic environmental conditions, it was expected, would turn these farms into what Gieryn calls a "truth spot," or a space for the production of "explicitly situated knowledge" that would become an undeniable scientific fact.[4]

However, as has been widely noted in the literature on the topic, the practice of science in these settings seldom follows this ideal model.[5] In clear contrast with laboratory practice, one pervasive characteristic of experimental farms has been "the embedding of experiments in a specific ecological, material and institutional environment."[6] This embeddedness has usually been the source of many problems, since the heterogeneous array of entities forming local ecologies, from animals and plants to visitors and scientific counterparts, usually challenges the paths laid out by the designers of these experimental systems.

Loncha station was not an exception to this trend. Since its inception it has experienced multiple upheavals and challenges, as La Pobrecita's suffering and the concerns of actors such as Venegas reveal. As a consequence, not only have the particular experimental practices, compounds, and results obtained been continually changed, but also the farm as a whole has been regularly reshaped, "being heaved up anew and then worn away."[7]

PRODUCING FACTS AT THE FARM

After several months of intense negotiations with its then owners, in December 1983 CODELCO was able to buy Hacienda Loncha. As seen in chapter 1, the pressure on the corporation was intense because Teniente's current tailings dam was about to reach its maximum quota, making the disposal of tailings in the near future highly uncertain. If they wanted the mine to remain fully operational the future dam should become operational soon, so the building process started right away.

An important issue at the time was the future status of Carén Creek. Although it only carried a small amount of water, it was extensively used for watering crops and animals by the small landowners downstream from Loncha, for most of them the only source of irrigation water. The first projections made by CODELCO indicated that a fully functional dam would release around 2.2 mt^3 of water per second throughout the year.[8] This regular flow would mean a massive increase in water availability for downstream landowners, running throughout the year, so it was quickly understood as a substantive improvement in terms of sheer volume.

The quality of that water, however, became more of a contentious issue. As related to us by Gustavo Quiñones, then head of the environmental area at Teniente: "When we announced the impact that the dam was going to have locally [in 1984], the strongest [feared] impact, the main worry, for the community, was the quality of water. . . . [It] caused high uncertainty on the zones located downstream." These fears were confirmed two years later when first tests on the water starting to accumulate inside the dam showed high concentrations of several chemical components, especially sulfates and molybdenum, surpassing Chile's legally permissible maximum limits for surface waters.

To properly explore the issue, in 1986 CODELCO hired CICA Ingenieros, an environmental engineering consultancy firm, whose initial results were worrying. Federico Bustos, one of the CICA directors, recalled that after running some tests, "I said to them: 'Look, this water is going to be gypsum-saturated'. I mean, they are going to have 1,800 parts per million of sulfates and more than 600 parts per million of calcium . . . and it is just a matter of time for them to join up and form gypsum, that is the stable form of sulfates." High gypsum concentrations in soils are usually reduced by constant watering, but that was not an option in an arid place such as the Carén basin, so it was expected that in the medium term gypsum would make local soils largely unproductive.

With these results in hand, Bustos went to the University of California Riverside, the location of the foremost experts at the time on these kinds of heavily intervened soils. They produced there "a very complex model, a thermodynamic one, and from this model emerged that gypsum would never form, that as long as you keep certain humidity, gypsum would never emerge." This result was encouraging for CODELCO, since it seemed to dismiss the initial fears that the water from the dam would have an overall negative productive effect on the fields downstream. Out of these first measurements and modeling exercises, the happy coexistence thesis started to emerge among CODELCO personnel and the consultants.

However, before this thesis could be further mobilized, especially to local farmers, more testing was needed. After all, the model also showed that salts and molybdenum in the water would increase substantially, well beyond the legally permissible maximums for the latter. For this reason, the proposal was to continue studying the issue at a local level through an experimental farm. Such a setting would allow, as Bustos said, running experiments on "a replica of the natural conditions," echoing general arguments about the advantages of in-situ experimental settings. However, the experimental farm's aim was never solely to produce certified knowledge. Obviously such knowledge was deemed valuable, allowing Teniente personnel to advance the coexistence thesis based not on a conceptual model made in California but on certified scientific knowledge emerging from a local replica of natural conditions. At the same time, the farm was also oriented from the very beginning to the "people downstream, small farmers, that needed the demonstration effect, I was counting on the participation of those people, for them to see that there's nothing fishy here and that we weren't running the experiments with water from wells." As noted by Bustos, the farm also looked to enact a "demonstration effect" for the relevant local—and increasingly nonlocal—public, seeking to convince them of the validity of the happy coexistence thesis through their direct experience on the site. As much as an "experimental system," the farm was also going to be a "spectacle," a living exemplar of mining and agriculture's mutually enriching relationship at the Carén basin.[9]

The station's objective was described as "establishing possible toxic effects on ovine, bovine and rabbit animals of consuming fodder produced on soils irrigated with tailings water and/or by the direct consumption of such water."[10] The main issue at hand was, then, how different living beings would react to the presence of the novel chemical compounds that the water from the dam would bring, especially molybdenum and sulfates. Given the predominantly agricultural uses of local lands, the reaction would be read from a conventional agricultural productionist frame of comparison, or "a typically linear orientation to the future based on producing output and profit through innovation," resulting in the exclusive study of conventional agricultural parameters of farm production.[11]

Following an experimental approach, the production of the data consisted of two interrelated operations: baselining and testing. The first involves the practices required to set a baseline, understood as the "documented or reconstructed historic state of being for a particular aspect of nature—an ecosystem, a place, a watershed."[12] Given the comparative valuation mode at play in most environmental science, establishing those baselines appeared to be a compulsory stage for any research activity, operating afterward as a contrasting point for any findings.[13] Following the scheme for experimental systems developed by Rheinberger, we could consider baselines the ultimate technical objects of environmental science without which no epistemic device could be properly perceived and/or measured.[14]

Baselining at Loncha consisted, first, of establishing "natural" weather and irrigation patterns for the area, so they could be maintained during the forthcoming experimental operations. Second, some regular local crops and animals were brought in to be used as the main "experimental organisms" for the study.[15] After making an inventory of local agriculture, the consultants settled on crops such as alfalfa, vegetables such as lettuce, various kinds of fruit trees, and livestock such as sheep and cows. Given the importance of having a valid comparative base, it was key that, at least initially, these organisms were highly representative of the area's agriculture. To secure representativeness, seeds, tree buds, and animals were bought from a public agricultural experimental station located in an area of similar geographical and productive characteristics as Carén.

Once those baseline conditions were considered acceptable, the proper field testing started. The experimental procedure for both plants and animals followed a similar pattern, as can be seen in the case of commercial rabbits, described in a summary about the main results of the research in the 1987–1993 period.[16]

> The objective of the study was to know the effect [on rabbits] of the consumption of forage irrigated with tailings water and its use for drinking, for which three treatments were designed:
>
> • Treatment 1: Food and drinking effect (food from Loncha + tailings water)
> • Treatment 2: Food effect (food from Loncha + well water)
> • Treatment 3: Drinking effect (external food + tailings water)

The effect of forage and drinking water was evaluated in relation to two productive variables (body weight and hair production), the latter being of interest in rabbit farms, and one reproductive variable (litter size at weaning).

The control group consisted on a commercial productive hatchery, located in the municipality of Peñaflor, Metropolitan Region. The animal type, sex, age, and reproductive management used as control is similar to the one used in this study. The information obtained was also compared with bibliographic sources.

Testing consisted mainly of the creation of different experimental entanglements between tailings water and the farm organisms, from plants such as alfalfa to multi-stomached animals such as cows. As explained in the preceding quote, these entanglements took three different forms: the direct intake of tailings water by the rabbits, their consumption of forage irrigated with tailings water, and a combination of both situations. The consequences of these encounters for the living entities were then measured and used as an indicator of the overall effect of the nascent geosymbioses between tailings water and these organisms. Notably, these effects were measured only in terms of productive variables such as weight, hair production, and the size of the litter at weaning. The baseline control group was composed of rabbits located in a faraway location, a commercial hatchery some 150 kilometers away. According to the conventional approach, these entanglements were understood as following linear patterns, leaving no space for any kind of messes or unexpected alliances.

From CODELCO's point of view, the initial assessments of the results of this experiment could not have been more positive. As a 1993 summary report from the consultants concludes, "The contents of sulfates and heavy metals entered into the edaphic profile during seven years of irrigation with tailings water at the Loncha Experimental Station allow normal growth, development and productivity of the different plant systems analyzed."[17] In the case of animals, the situation was also encouraging given that the experiments "show[ed] the feasibility of feeding polygastric cattle with forage irrigated with tailings water and obtain[ing] normal production parameters."[18] Specifically, as stated in a later presentation, "sulfates and molybdates are not phytotoxic ions and its current limitation in the regulation for irrigation water correspond to [the local adoption of] foreign standards with national and institutional characteristics quite different from Chile."[19] In sum, there was no problem with sulfates or molybdenum in tailings water as such, only with the Chilean regulation, which was too strict and did not recognize these chemicals' innocuousness.

Instead of being toxic, the geosymbioses emerging from the use of tailings water for agriculture were at least commensal, if not mutualistic. As stated in the CICA report, "given the water scarcity that characterizes this valley, the agricultural use of effluent [tailings] water represents a potentially interesting agroeconomic benefit for the area, allowing the irrigation [of what would otherwise be] dry land."[20] Happy productivist coexistence was not wishful thinking on the part of some of CODELCO's executives, but a scientific fact, the report suggested.

Such a reading of the research being done at Loncha was not shared by everyone, though. In the same year CICA published its report, the Dirección General de Aguas (DGA), Chile's national water authority, published a parallel assessment of the research carried out at the Loncha Station that went exactly in the opposite direction.[21] Written by academics from Universidad de Chile, it reviews CICA's technical reports, finding a number of highly problematic issues. The assessment especially criticizes the lack of proper control groups in the form of plants and animals located on the farm but that receive neither food nor water containing tailings. Adapting from Ankeny and Leonelli, it was argued that CICA research initiatives lacked valid *baseline organisms*, or experimental organisms that could be taken as the standard agricultural organisms for that specific area.[22] As the report states, "No baseline is established in terms of chemical analyses (previous levels of heavy metals in plants, animals, etc.), necessary to estimate the impact of the increase of some chemical species in the ecosystem."[23]

Following Rheinberger's model, an agricultural scientific study without proper baselines cannot really identify whether an object of research has true epistemic properties, any potential epistemic thing "not becom[ing] shaped, but . . . rather dissipat[ing] in the hands of the researcher."[24] As a consequence, the plants and animals being tested at Carén ended up behaving the way the actors who designed the experimental system wanted them to behave. For this reason, it was not strange that they simply confirmed the assumptions made by CICA personnel (and their counterparts at CODELCO) from the very beginning. The DGA report concludes with the damning affirmation that "the proposed conclusions are not relevant to the results actually obtained. Rather, they correspond to statements based on literature. . . . For the above reasons, it is not possible to draw conclusions about the use of tailings water for the production of forages or the drinking of animals."[25] These openly critical assessments from a respected regulatory institution meant that the data produced in the first decade or so of experimental research at the farm could not be considered proper scientific facts.[26]

As a result of the difficulties found in turning the happy coexistence thesis into a proper scientific fact, from the mid-1990s the station was used mostly as an educational space where children from nearby locations could learn about the work in a mine such as Teniente and the efforts made by CODELCO to protect the environment. This usage was so prominent that Venegas candidly remarked to us that "the farm has never been an experimental farm, it was a demonstrative farm. . . . [I]t was not an experimental center." Its past as a formal experimental research station—the decades of research and results, the massive investments, the technical reports and papers that were never published—all could be forgotten. Education was the only business in Loncha, or so it seemed.

MINERAL PARASITISM

The events of April 16, 2006, involved much more than the spill of several hundred thousand tons of tailings downstream from the dam. They also reactivated the public discussion regarding tailings' potential toxic effects on the surrounding environment, which had been dormant for years. To deal with them, Loncha's former incarnation as an experimental research station was hastily resurrected, albeit with some key differences from its former setup. As seen previously, the experimental station's original motivation was mostly CODELCO's internal doubts about the potential negative effects of tailings water on highly conventional agricultural practices. The reincarnation of the station in 2006, in contrast, emerged after massive amounts of tailings were already deposited in the Carén creekbed. To make matters worse, there were several accusations regarding the death of animals due to the intake of water carrying tailings, one resulting in an extensive lawsuit (explored in chapter 5). As a consequence, the spill raised great levels of public alarm, covered frequently in the media and even motivating the creation of an investigative commission by Chile's national parliament to establish its causes and possible consequences.

Given such framing, now there was a pressing need to produce solid scientific evidence about tailings water's toxic capacities, proof that could be mobilized on the multiple fronts that CODELCO was facing at the time.

Learning from the problems faced in the past, the new team from CICA in charge of running the scheme opted for making some substantive changes to the previous experimental design. This was done especially by constituting suitable baselines in terms of contextual factors (water, climate, etc.) and, probably more important, by having proper baseline organisms: crops and animals located at the farm that had no contact whatsoever with tailings water.

As soon as these baseline organisms were introduced, the geosymbiotic entanglements between tailings water and plants and animals finally became proper epistemic things. However, to the dismay of CODELCO personnel, they did not follow the path toward happy coexistence. On the contrary, as recalled by Gerardo Manríquez, a key member of the new team from CICA, "Our results determined very different things from what we have been seeing historically. . . . Indeed, we had serious restrictions [in multiple parameters]." Crops and livestock at Loncha were not entangled in mutualist or even a commensal geosymbiotic entanglements with tailings water, but rather in parasitic and even toxic ones.

Regarding crops, Manríquez noted that "the results were super clear. In the control plot, which had never been watered [with tailings water], we had huge clovers, huge bulbs, we had to cut them every week, in the other, definitely, first, the germination was very open, and [second] the growth rate was undoubtedly much lower." Instead of being merely innocuous presences, inside the plants these chemicals became parasites, damaging agents clearly weakening their vegetal hosts. After entering into geosymbiotic entanglements with tailings water, plants showed significant effects, mutations, and lower growth rates.

To the concern of CODELCO, those detrimental effects were not only evident in crops but also, and even more intensively, in farm animals. After introducing baseline animals in the form of sheep and cows who never drink tailings water or ate forage irrigated with it such as Matilda, "the differences were obvious," as Manríquez stated. On the one hand, "the group [of sheep] that ate [external food] and drank water from outside was a bedtime-story group, one of those that you count to fall asleep." On the other hand, the group that drank water from Carén Creek and ate forage irrigated with it "were sheep that passed with diarrhea, and a sheep with diarrhea is rare, sheep shits like a rabbit!" Openly contradicting the results of the previous studies, the sheep appeared "undoubtedly severely affected" by tailings water, experiencing an important loss of weight and wool. In the case of cows, the negative effects were so strong that the animals refused to drink tailings water altogether during summertime, even putting their lives at risk, as happened with La Pobrecita.[27] In both cases, tailings' chemicals behaved as damaging agents in their animal hosts, not only affecting their growth but even threatening to kill them altogether, a clearly toxic kind of geosymbiosis.

In all, the clear differences between baseline and epistemic organisms in the second wave of experimentation at the station confirmed the initial fears regarding the detrimental effects on plants and animals upon entering into geosymbiotic entanglements with tailings' chemicals such as molybdenum and sulfates, even adding some new concerns.[28] In almost every relevant parameter the research found worrying trends, a finding quite far from the happy coexistence thesis. Parasitism was the rule, an entanglement that allowed plants and animals to keep living and growing, but significantly affected their development and long-term survival. Toxic entanglements were also prominent in the case of cows.

The prominence of parasitic entanglements between these organisms and minerals does not necessarily mean that agriculture was doomed in the Carén Creek basin. Although the tailings water weakens them, hosts can have long and fruitful lives while engaged in parasitic entanglements with it. In fact, all of us regularly host various kinds of parasites, forming a normal kind of relationship in most ecologies.[29] Regarding toxic geosymbioses the situation was clearly damaging, canceling most possibilities for agriculture in the valley if they became widespread. But if such situation could be avoided, certainly some kinds of agriculture could be practiced.

Implicitly agreeing with this notion, the consultants advised CODELCO to assume "its responsibility regarding the [chemical] concentrations found downstream," as recalled by Luisa Figueroa, an agronomist from CICA. Starting by taking regular measurements of several chemicals' concentrations in animals and crops all along the valley (not only on the farm) and taking a census of the kind of agriculture actually practiced in the basin, they urged the corporation to develop a fully alternative approach to the issue. As they argued, up to that point "the experimental station was working . . . [with the] aim of seeing its evolution in terms of contamination, but was not seeing the issue of the best measures applicable to an already contaminated system, so we proposed them, to turn over the research a bit to see how to manage an already contaminated agricultural site, in order to identify the best possible agronomic practices." At its core, CICA's recommendation asked CODELCO to fully embrace the fact that the dam was having a detrimental effect on local agricultural beings, renouncing in the process the notion that agriculture as usual could be practiced in the Carén Creek basin.

In its place, the consultants proposed, CODELCO should accept the notion that the coexistence between tailings' chemicals and living entities at Carén was not necessarily good, but also not inevitably bad. These geosymbioses were ambivalent, continually changing, sometimes benefiting all the partners, sometimes damaging some of them. So the farm should be opened up as a site for public experimentation with geosymbioses, inviting CODELCO personnel, experts, and neighbors to openly discuss the issue, or, in Figueroa's words, "to say 'such a practice served me', 'to cultivate forage with such and such practice', 'with such and such fertilizer', 'with such Ph management', I don't know, whatever it is, 'it works in such a way.'" Instead of only making diagnoses aimed at showing the lack of negative effects, Loncha as a site for public experimentation would aim at developing tools to support current local farming practices. This would mean transforming the station into a place to engage with *parasitic agriculture*, or an agriculture specifically built to adapt, even to flourish, in structurally polluted landscapes such as Carén.

ETERNAL EXPERIMENTATION

Matilda, La Pobrecita, and their two nameless companions were subjects of a new round of testing at the experimental station, started in 2014. Although another consultancy company was in charge this time, the objectives of the initiative were strikingly similar to those of the very first project carried out by CICA in the mid-1980s: "to generate a diagnosis that allows to identify and evaluate the environmental impacts of tailings water on ... [local agricultural] ecosystems." Such similitude was not accidental; it was the direct result of the main strategy that CODELCO had adopted regarding the station and its "facts."

The contours of this strategy were presented to us in July 2014 by Héctor Galindo, a high-level director from the environmental area at Teniente, when discussing the possible steps to take based on the results from the experimental station. First, he said, the strategy would involve making the issue public, "but in a very intelligent way, you should not shoot yourself on the foot, you should be quite [careful]. . . . At some point, this information, these results, will have to be reviewed here, internally, first, [and then] the pertinence of informing the community [should be discussed]." After the information was made public, they wanted to start "working with the farmers, because in the end you cannot leave these people just there, doing nothing." Up to this point, the strategy sounded quite similar to the one proposed by CICA in the aftermath of the 2006 spill, to turn Loncha into a site for public experimentation in parasitic (even toxic) agriculture.

However, there was still one key limitation to launching this new incarnation. Before doing anything, Galindo continued, "We need more concrete results. Today I have only some notions, I have some results, but not enough yet as to be able to communicate them." After more than three decades of almost continual research at the experimental station, after hundreds of experiments and dozens of reports, the knowledge obtained was dismissed as merely "some notions." CODELCO's directors were still waiting for more "concrete results" before recognizing their responsibility and implementing measures to deal with it. They just needed a bit more research, a few more tests on plants and animals, a few more reports. Then they would be able to act. But only then.

Departing from the usual strategy of denying the scientific validity of results going against corporate interests, in Galindo's words we can see a more sophisticated form of denialism that we call *eternal experimentation*. This strategy is based on the never-ending demand for more and more evidence before any suitable action can be taken on an issue, a demand that has succeeded in making sure that science regarding the dam's ultimate effects on the local agricultural systems remains eternally "undone," so no proper responsibility, acknowledgment, and/or corrective actions can be implemented.[30]

The strategy was born, mainly, from the recognition that although full of possibilities, the path toward accepting parasitic/toxic geosymbioses proposed by CICA in 2011 entailed important risks. Chief among them was that it involved CODELCO publicly recognizing that the company had been heavily polluting Carén basin for decades and, probably more critical, would continue to do so well into the future, even decades after the dam officially closes. Therefore, this would imply making an argument about the deep time of mining-related chemical pollution, departing from the usual narrative of "natural" geological sedimentation as the only possible future for areas such as Carén. At the same time, it would also mean discarding the happy coexistence thesis and its conventional productivist frame, accepting that any agriculture to be practiced in the basin would be of a *parasitic* kind, a deeply affected endeavor, both detrimental and vital, always on the verge of becoming toxic.

In all, to follow this path would imply accepting residualism as a largely failed project at Carén, neither a geology (sedimentation) nor a biology (happy coexistence). In contrast, enacting further research schemes to try to finally materialize happy coexistence appeared to be the less radical alternative, even if it meant subjecting more nameless cows such as La Pobrecita to painful experimental conditions.

The failure to produce data confirming the happy coexistence thesis, however, does not mean that locals beyond the Carén complex have remained oblivious to the kind of parasitic agriculture that could be practiced in the basin. On the contrary, although CODELCO has so far resisted calls to start understanding/diffusing this parasitic agriculture at the experimental station, locals have been busy practicing it for decades. For them, coexistence with tailings water has been a fact, but it is rarely acknowledged as "happy," as explored in the next chapter.

FOUR

Parasitism

REGULATORY PURIFICATION

A fundamental change occurred to tailings water during its movement between the PAMo (chapter 2) and its discharge point on Carén Creek. This change was not only a matter of chemistry, of a certain reduction in the concentrations of molybdenum and other minerals in the water. It was also a matter of valuation, of its legal status and the qualities and regulations attached to it.

When water emerges from PAMo it is still a *residuos industriales líquidos* (liquid industrial residues or RILS), using Chile's technical denomination for industrial wastewater.[1] In a similar way to most regulations on the matter, RILs are seen as potentially harmful chemical substances whose release must be (albeit always imperfectly) controlled, especially regarding their toxic capacities. This conceptualization implies that RILs always have a negative identity, "that its useful life is effectively over; that is has aged, moved on, or is otherwise to be disregarded."[2] RILs are troublesome residues that must be neutralized so that no harm comes to any living entity, especially human beings. With this aim in mind, the Chilean regulation establishes certain maximum concentrations of several chemicals that RILs cannot surpass when being released into a water stream. Through a process of technical certification, in the form of an independent lab continuously taking samples from the Carén Creek and uploading the resulting data to the water authority's website, water-as-RILs is certified as able to be released.[3] In most cases, this certification means an end to environmental concerns related to such water on the part of CODELCO. As Héctor Galindo from the environmental department of Teniente joyfully declared to us, after the external certification "you can say 'I fulfilled my environmental commitments!'"

Quite paradoxically, as soon as this water loses its status as residue it becomes its direct opposite: a commodity. In Chile, water is governed through an instrument called the Water Code. Enacted in 1981, in the midst of the neoliberal revolution carried out by Pinochet's military dictatorship, the code enacts water as just another kind of private commodity. Private actors can acquire water rights and freely trade them, theoretically allowing more rational and efficient water allocation. In practice, however, the Water Code ended up concentrating water rights in the hands of big corporations and speculators, leaving most small farmers and indigenous communities with very limited access to water, even for human consumption.[4]

The monetary value of water coming from the dam has been significantly enhanced in recent years by the megadrought affecting central Chile, making any water available precious. However, from the standpoint of CODELCO, there was a risk if water rights were bought by local farmers in order to irrigate their lands and animals. As seen in the previous chapter, after decades of testing at the experimental farm, CODELCO was well aware that this water might cause severe negative effects when being used for agricultural activities. Albeit no longer technically a residue, it could still be a toxicant.

To the good fortune of CODELCO, another actor had an almost complete monopoly on the rights over these waters. Twenty kilometers downstream from the dam lies Rapel, a hydroelectric plant inaugurated in 1968 by Empresa Nacional de Electricidad Sociedad Anónima (ENDESA). For this corporation, water quality was absolutely irrelevant. All that mattered was volume, the sheer amount of water available for propelling Rapel's turbines. A mutually beneficial arrangement was rapidly agreed, in which CODELCO released the water at Carén and ENDESA appropriated it, given its ownership of most water rights in the area. After being treated in the PAMo, tailings water was to be released into Carén Creek, starting a 20-kilometer trip to Rapel Lake, the water reservoir for the hydroelectric plant. At the lake, Carén's water would mix with that coming from other sources, chemically and regulatorily dissolving any responsibility that CODELCO could have regarding its potential toxicity. Once the water reached Rapel, all was fine. The problem lay in the space between the two points.

As can be seen in figure 14, on its way to Rapel, Carén Creek passes through several small villages, such as Pincha and Quilamuta, home of low-income farmers mostly dedicated to subsistence agriculture. The problem

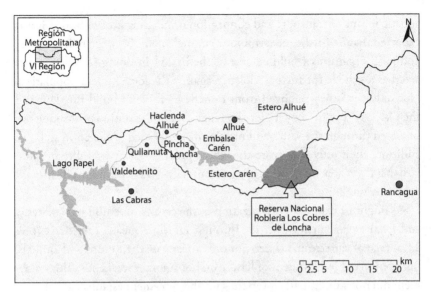

FIGURE 14. Villages between Carén Dam and Rapel Lake

from the standpoint of CODELCO lay in the fact that they had historically used the water of the stream to irrigate their crops and the animals, having in most cases no alternative source. Therefore, it was highly likely that they would continue to do so, with all the potential negative effects outlined in the previous chapter.

The farmers' lack of water rights appeared to be the best possible way to restrain them from extracting water from the creek. After all, if they insisted on doing so, they were liable to be legally prosecuted for theft of private property. Even if this last option was never really implemented, the privatization of the stream had a more practical benefit for CODELCO: it absolved the company from legal responsibilities in the case that the consumption of the water ended up causing any harm to crops, animals, or humans. They were not meant to be using it, so they could not complain if its usage caused any trouble. In theory, they could even be prosecuted for becoming contaminated.

Water commodification not only turns a residue into a valued good but also criminalizes its consumption, exempting CODELCO from any responsibility in the matter. Given this setup, locals were expected to leave the water alone, without engaging with it in any way, performing a final turn in the residualist loop. Through a regulatory removal of the "waterpath", the crucial aquatic means of transport and connection originating geosymbioses, it was expected that no further geosymbiotic entanglements between tailings water and local organisms would emerge at the basin.[5] In doing so, water rights acted as a vehicle for further violence against the local human population. Not only were they removed from Loncha by force to build the dam and then "slowly" polluted by under-the-radar chemicals continually leaking out of Carén through the water and massive disasters such as the 2006 spill.[6] In addition, their only gain from the whole process—the possibility of regularly having water for their crops and animals—was criminalized through privatization.

As might be expected, privatization of the creek's water did not suddenly end local engagements with it. Through multiple means the locals took advantage of the stream's water, not only to irrigate their crops and animals, but even to enjoy a moment of leisure on hot summer weekends. This enjoyment did not substantially interfere with the agreement established between CODELCO and ENDESA—most of the water still arrived at Rapel—but it further eroded the residualist claim that happy coexistence was the only way in which locals could engage with tailings water.

INVERSE RESISTANCE

As soon as we started talking about Carén Creek water on our first visit in October 2014, Enrique Concha showed us the sores in his hands (see figure 15). They were little red spots, with a diameter of around one centimeter. He attributed them to the presence of "acids" in the water, regularly appearing after he had spent some time watering his crops. For him pollution was not only something external, something experienced when seeing strange white spots on the soil after watering his field or when some animals refused to drink water from the stream. The water's pollution was personal, intimate.

Concha lives in Pincha, around 5 kilometers downstream from the dam. Like many of his neighbors, he has a small plot of land where he plants potatoes, which, once harvested, are sold in a vegetable market in Santiago. Twice a week he goes down to the creekbank at seven o'clock to prepare his water pump to send water to his crops (see figure 16). After turning on the machine and checking that there are no filtrations, he immediately returns to his plot and starts canalizing the water through several furrows filled with hundreds of potatoes. Commonly, this task takes him an entire day.

FIGURE 15. Sores in Enrique Concha's hands, 2014

FIGURE 16. Enrique Concha's water pump

While describing his routine to us, Concha spontaneously commented on several situations that, in his opinion, indicated that Carén Creek's water was severely polluted. Besides showing us his hands, he told us, for example, that after irrigating his crops a crust appeared on the soil, "a crust as white as salt." The same phenomenon occurred with the water pump, which was covered with a notorious tartar that, according to him, rusted the machine. "This is awful," he repeated over and over again while recalling these issues.

Although these descriptions varied from case to case, all farmers in the area identified tailings at the dam as the source of the pollution. They knew about the potentially toxic chemicals the water carried from the dam to their lands and animals, even into their bodies. They all knew that. But even so, they did not appear interested at all in getting involved in a "water war" to correct this situation. Even Julio Casas, another farmer from Loncha who compared CODELCO's arrival in the area with the arrival of a "pest," showed a certain restraint when criticizing the corporation. After all, he knew that CODELCO "controls the amount of water [in Carén Creek], so you have the same amount of water now, in October, and in December, January, February, [while] in the past it was different [since] in this season there was less water than now, and in March there was only a little bit of water, only some puddles that rapidly dried out." Thanks to the dam the farmers were less vulnerable to seasonal changes, having always more than enough water to irrigate their crops. This regularity was especially valuable when considering the ongoing megadrought, which has been especially damaging for low-income farmers.[7] Hence the regular availability of water in Carén Creek was not something to be dismissed. Given that around 90 percent of the water in the stream comes from the dam, Carén Creek is probably the only creek in the whole central valley of Chile relatively safe from the effects of the drought.[8] As a consequence, it has become the ultimate lifeline for farmers of the area, the difference between having a means of survival or not. As Ricardo Monsalve, a farmer from Pincha, said to us, "I have no alternative, I mean, [without water] my land is useless."

To use this lifeline, though, they have had to overcome the double circuit of dispossession posed by water's privatization and pollution. Given their very limited resources and the strength of their opponents—two of Chile's largest corporations, supported by the law and the state—they simply could not engage in a formal environmental justice movement demanding legal access to the water and/or its proper cleanup. In contrast, their forms of action were much closer to what Papadopoulos calls "more-than-social-movements" or forms of resistance that "do not attempt to contest power by organizing protest; rather, they attempt to create the conditions for the articulation of alternative imaginaries and alternative practices that bypass instituted power and generate alternative modes of existence."[9]

In the villages that lie between Carén and Rapel, on dozens of small plots of land, farmers resisted the attempts to exclude them from using this water by engaging in an alternative form of agriculture, one that could flourish *without* water rights and *without* environmental purity. This move implied enacting what we have called a *parasitic agriculture*, or a form of agriculture highly adapted to flourishing in a permanently polluted and privatized environment. This parasitism was not only environmental but also political, in the sense of dismissing the regulatory bodies calling on farmers to refrain from using the water passing in front of their fields, opening the door to alternative modes of relating to water.

At first sight, parasitic agriculture is a disposition to simply disobey calls to consider water as a private good, to refrain from using it if one does not have rights. This disposition starts by recognizing the existence of water rights and, at the same time, dismissing such rights as utterly unjust. As expressed by Monsalve, "I have heard that the owner of the water is ENDESA since [it controls] the [Rapel] lake. . . . This is the thing; the bigger ones always control water. I do not know why ENDESA controls always all the water available in the country!" One of the most controversial features of the Chilean Water Code is the separation it introduces between water and land, treating them as commodities that can be traded separately. As a consequence, water ends up being traded in several private water rights markets, which do not take into consideration the inhabitants of the basin whose water resources are at stake, especially low-income farmers.[10] They are utterly invisible to the system and, hence, the water from a basin could be entirely allocated to the higher bidder, in this case ENDESA, without any consideration to who is using it already.

In clear opposition to this setup, Monsalve establishes a clear connection between land tenure and access to water. As he claims, "My land does not serve [if I cannot water it], regardless if I have water rights or not. I mean, we have used [water from the creek] since my father bought [this plot], and, before he bought it, the owner of this field took also water from the creek to irrigate. So, I will keep on using it, I have to keep on using it. . . . I do not have an alternative." In these words, Monsalve practically asserts that before (or besides) private water rights are *riparian rights*. Legally recognized in countries such as Canada and Australia, riparian rights refer to a system in which "owners of land abutting watercourses have a right to reasonable use of the water."[11] In contrast with private water rights such as those existing in Chile, riparian rights "do not originate with the state; they evolve out of a given ecological context of human existence."[12] Riparian rights are not property rights, but a right to enact a relationship with a certain watercourse for those living near it. It is a right of use, not appropriation. Parasiting on the private water scheme, Monsalve and his neighbors restored by practice traditional riparian rights, allowing them to keep practicing agriculture amid the devastation caused by the megadrought.

However, the water they were using was not the same as their parents had used. Even ignoring the results of the experiments at Loncha Station, farmers were conscious that nothing like a happy coexistence existed in the Carén Creek basin. The water carried something, some sort of unknown pollutant, that produced strange effects in plants, animals, and even them, as the sores on Concha's hands testify. These are polluted waters, pollution that comes from extraction. Consequently, the farmers have developed multiple practices aimed at engaging with this water, techniques for bringing tailings water to their farms while keeping their crops growing and their animals (and even themselves) healthy.

This process starts by acquiring knowledge, in multiple forms. Despite having only a few years of formal education, Monsalve talked to us about how "some studies say that . . . minerals accumulate in the soil over time, I mean, the productivity of the soil starts to decrease." This knowledge came from many sources, including informal conversations with people involved in the experiments carried out at Loncha.[13] Some knowledge was of a general nature, but some presented a high degree of sophistication, with Monsalve becoming an amateur soil scientist: "Because of tailings soil's electrical conductivity is higher, electrical conductivity increases when the soil has a great number of minerals, and when the electrical conductivity is high plants cannot absorb nutrients. For this reason, one should use guano, because plants cannot absorb what they need." This knowledge translated into the incorporation of several items into local soils, such as guano and lime, with the objective of turning the geosymbioses from overly toxic toward parasitic or even commensal.

In parallel with applying external knowledge, Monsalve has also developed his own experimental systems to explore new ways to engage with the water of the creek (see figure 17), initiatives that he calls his *proyectos*, projects, as he explained one afternoon in April 2015:

MONSALVE: This is a belt used for drip irrigation. [Showing the belt] Here you have the droppers. . . . [T]his one has 10 [droppers]. There are belts with 10, 15 and 20 [droppers]. A couple of years ago, I planted some potatoes and made a test with these [irrigation belts], I wanted to see if they worked. But the water from the creek does not work since tailings block the droppers. So, I made an experiment with water taken from a well. And the productivity was spectacular. . . . [T]he total productivity

FIGURE 17. Ricardo Monsalve watering crops

grew from, for example, 500 sacks per hectare to 700, something like that . . .

FLORES: And you say that tailings block the droppers . . .

MONSALVE: When you irrigate with water from the creek, tailings block the droppers. For this reason, when I bought this hose, my idea was connecting it to a 4 inch-pump, but using a filter in the pump [to catch tailings particles].

In contrast with the experiments carried out at the farm, this one was not focused solely on obtaining knowledge but mostly on identifying concrete solutions, resembling CICA's last proposal. As his experiment revealed that tailings in the water tended to block the droppers on the irrigation belt, Monsalve decided to add a filter to the hose taking water from the creek in order to remove some of the particles blocking the droppers. In doing so, Monsalve did not seek to get completely pure water, such as that his father probably used before the arrival of the dam. Despite the experiments showing him that such water would substantially increase his productivity, he was well aware that it was well above his means. Accordingly, he opted for a more pragmatic approach, aiming to reduce somewhat the concentration of tailings in the water, so his land would continue to be productive in the short term.

In other cases, the best strategy was simply to withdraw from taking any course of action, leaving those who were going to be affected by the water with the decision of whether to engage with it. This happened in the case of Julio Casas, whose horses and mules had clear routines for engaging with the creek's water during the day. In the morning, "I take them to the creek and they do not drink water. . . . I offer them water [from the creek], they do not accept it, so I take them to drink tap water, I give them water from the tap." According to Casas, the reason for this behavior was that at the beginning of the day, water "smells like tailings." In the evenings, when the water in the creek did not smell like tailings, the animals had no problem drinking it. As with La Pobrecita in the previous chapter, the changing perceived status of the creek required different watering tactics, with the aim of balancing the animals' well-being with reducing the higher expense involved in watering them only from the tap.

By implementing these practices, the farmers were not only able to keep their fields productive, but also were consciously departing from the ideal standards of purity behind most Chilean regulations regarding agricultural irrigation water. We were able to witness this process one afternoon in November 2014 when again visiting Enrique Concha's farm. While we were looking at his crops, he commented, "Imagine what would happen if people from sanitary [agencies] come here?," suggesting concern about being sanctioned for irrigating with water from the creek. As already seen, in Chile surface water must satisfy several quality criteria in order to be used for irrigation and watering animals. In particular, Norm 1333, enacted by the Instituto Nacional de Normalización (National Institute for Normalization) in 1978, defines several chemical concentrations thresholds that could not be surpassed by an irrigation body of water, mostly based on international standards.

It was immediately obvious that a strict application of this regulation would mean the end of agriculture in the Carén basin, especially for small producers such as Concha, who do not have resources to pay for cleaner sources of water (such as deep wells). Their only option is to knowingly ignore such regulations, hoping that no one from the DGA or any environmental health institution would come to oversee the matter. In this refusal to abide by the rules, they have been greatly helped by CODELCO's turning a blind eye to the issue, mainly given the high cost (both economic and in terms of community relations) of continually prosecuting anyone using the creek's water.

This implicit acceptance was not unproblematic for the corporation, as recalled by Héctor Galindo who commented that "everyone uses the water, it is a very high consumption, in fact, we made an hydrogeological model that established which are the sectors [where consumption is higher], there are many inlets and pumps through which the people takes water out, it is a matter to stand at [the Carén complex] entrance and you can hear the sound of the people's pumps extracting water, this is very well known." This extended usage put CODELCO in a tricky position. In legal terms, by discharging water that has mineral concentrations below the maximum for RIL, CODELCO has no responsibility whatsoever for what happens with the water afterward. As soon as the water crosses the Carén complex's barriers, it becomes someone else's problem, either the DGA's or ENDESA's, as owners of the corresponding water rights. But in practice the situation was more complex. As Galindo acknowledged, "You can debate whether you [CODELCO] are or are not responsible for that, in certain way you are not, because we discharge [in agreement with RIL regulations] and I am the owner of this property, but in other sense we do [have certain responsibility] because we discharge and they are consuming it, then to which degree I am responsible for that? And this is relevant, because at the end, from my perspective, we are responsible for what happens downstream." Beyond issues of property rights, the water was ultimately produced by CODELCO, so the corporation was going to be blamed for any negative effects derived from its usage by local farmers. However, the still open question regarding the ultimate effects of such usage, as explored in the previous chapter, meant that no proper definition of the issue existed. Eternal experimentation translated into eternal irresolution.

In the end, whether through practice (from the farmers) or inaction (from CODELCO), parasitic agriculture involved challenging legal frameworks stating that agriculture can be carried out only with proper water rights and under relatively pure natural conditions. If agriculture was meant to continue existing in the midst of a landscape hit by the megadrought, pollution, and privatization, it would do so by accepting that something like riparian water rights existed and that a degree of pollution was unavoidable, that at Carén purity is incompatible with geosymbiosis. In doing so, parasitic agriculture took the form of an inverse resistance, based not on avoiding any contact with polluted water or asking for its remediation, but rather on engaging with it in unexpected ways, forming strange alliances, engagements that usually meant subverting water's current legal and sanitary status.

It was a hot summer day in January 2016. We were on an embankment of Carén Creek near the town of Pincha, a site popular among people from nearby locations looking for a nice spot to spend a weekend in summertime. After parking our car, we looked for a place among the families who were enjoying picnics and barbecues by the creekside. The adults ate and drank, joyfully speaking with each other or just trying to take a nap. Meanwhile, children and teenagers plunged into the creek, either from trees or from an improvised trampoline composed of a polypropylene sack filled with sand. All this happened to the mixed beat of reggaeton music emerging from several loudspeakers.

Shortly after finding an empty place to sit, we started to feel the effects of the heat, which reached around 35 degrees Celsius that day. We were sweating, our bodies sticky inside our clothes. The few places protected from the sun were already occupied, so our only escape from the heat would be to plunge into Carén Creek. Just a few meters from us, we could see the gently flowing stream, almost transparent and surrounded by trees. Although we were there to do fieldwork, maybe taking a brief dip was not a bad idea. After all, this was what ethnographic fieldwork was all about, to use our bodies to experience life as close as possible to our subjects. To jump into the creek would then be of ethnographic value, besides relieving some of the heat. But we hesitated. We could not forget that, in spite of its nice appearance, the water flowing in front of us came from the Carén dam, it came from tailings, and we knew a lot about its high concentrations of some chemicals, some of them potentially toxic. However, as figure 18 illustrates, no one else seemed to share our hesitation.

Seated right next to us was Rodolfo Márquez, with whom we started chatting. That day he was picnicking with his wife, mother-in-law, and four children. He knew about the provenance of the water, but said, "I have come here since a long time ago, since my childhood, and I have never had any problem, never has happened something strange, . . . I have never arrived at home with marks [on the body], with . . . I do not know, when the skin turns into red and itch." Like Rodolfo, many of the people present on the creekbank that day had a long history of visits to the place, even in some cases stretching back to their childhoods. In recent years, given the megadrought, this spot had become even more popular, being one of the few places in the region at which they could temporarily escape the summer heat, with temperatures regularly surpassing 30 degrees Celsius in the afternoon.

FIGURE 18. People taking a dip in the Carén Creek

The availability of water was not the only reason the creekbank was so popular among the local population, though. The creek not only had water, but it was also open to everyone for free, allowing them to enjoy it without more expense than the gas used to drive there. This openness was also a rarity in Chile, where access to bodies of water such as lakes and reservoirs has been strictly regulated for decades, usually for the benefit of the second homes of middle-class people. This exclusionary access was especially strong at the nearest body of water, Rapel Lake. Less than a decade after its creation, in 1977, a regulating plan for the lake was enacted by the regional authorities, focused "almost exclusively on ordering leisure homes, . . . [housing characterized by] a strong functional and social segregation."[14] During its more than thirty years of functioning, this plan has caused an acute "lack of public places [by the lakeside] . . . producing important levels of segregation," as concluded in an official assessment of the matter by the Ministerio de Vivienda y Urbanismo (Ministry of Housing and Urbanism).[15] As a consequence, most of the bathing areas at Rapel were out of reach for local people such as Márquez. In this regard, going to the Carén creekbank had become another form of parasitism, in this time of regulations stating that only wealthy homeowners have rights to access the shores of lakes and rivers.

This practice was importantly helped by the extended perception among the visitors that Carén Creek's water was harmless. Although they were well aware of its origin, in all the years bathing at the creek nothing "strange" had happened to them or their families, a strangeness associated, in Márquez's case, mostly with the presence of skin lesions. This perception was even stronger when comparing Carén Creek with Rapel Lake, as noted by Nelson Fuentes, another person who was also picnicking by the creekside on one of our visits. After an absence of more than thirty years, a few months earlier he had visited Rapel again, but his experience of the place was not pleasant at all; upon entering the water his "first sensation was not that [he] was bathing in mud, but rather in pig shit!" This sensorial impression appears confirmed by the regular recurrence of pollution episodes at the lake, such as massive fish die-offs, mainly from the high concentrations of chemical residues brought to the lake by its tributaries.[16]

In contrast, Carén Creek appears somewhat more trustworthy. This is not only a matter of its water usually looking cleaner, relatively transparent, and without foul smells. Beyond this, the confidence in the water is based on fact that, as Fuentes tellingly declares, "I know what CODELCO does to lower the pollution. . . . I prefer taking a dip in this water, I know what it carries, the only recommendation is not drinking it, right? But in the other one I cannot take a dip because of the smell." As he acknowledges, this is not conventionally pure water, but neither is it the foul water at Rapel, whose origins are multiple and rather unknown. This is water whose pollutants are "known" to a certain degree, they come from the tailings deposited at Carén dam, and such traceability makes it somewhat more trustable.

This trust was clearly connected, as happened for Márquez, with the fact that the tailings were owned by CODELCO, whose environmental management was perceived as being in a constant process of improvement. As Fuentes continued, "Before . . . the water did not receive any treatment. It fell into the [discharge] tower because of gravity, and then it was released into the creek without being treated. But today there is more concern [from CODELCO]. They feel more concern about issues related to ph, turbidity. . . . They are concerned about several chemical factors that affect the water. . . . I don't know if it is cleaner, but it has lower tailings levels." Like several other local people, Fuentes appears to be quite well informed about the doings of CODELCO at Carén regarding water, even naming several of the water quality indicators that the corporation's personnel daily measured in the basin. This information appeared to be a clear indication that the corporation was taking the issue of water quality more seriously, especially the one released into Carén Creek.

Visitors such as Márquez and Fuentes were conscious that the water was not completely clean, but they believed that it was *clean enough*. Some chemicals were still there, they were certain. No matter the sophistication of the control systems implemented by CODELCO, as Fuentes suggested during our conversation, tailings cannot be completely removed from the water. But the lack of visible negative sensations and effects, and the constant monitoring on the part of the corporation, signaled to them that its concentrations were not really harmful. The creek water was certainly *impure*, but not impure enough to keep them from swimming on a hot summer day.

This perception, however, did not mean that they took the issue lightly. On the contrary, the risk of possible harmful consequences from swimming in the creek was always present, and consequently, visitors performed simple actions to protect themselves from the suspected chemicals in the water. Against conventional regulatory approaches to the issue—that identify the existence of multiple pathways for toxic exposure—for most of the people we encountered the only thing that really mattered was to avoid the accidental ingestion of water.[17] This precaution was applied especially to little children, as we were able to see in January 2016.

After being there for half an hour, a middle-aged man arrived on the bank of Carén Creek with a small child. Another kid, upon seeing them, left the bank and ran toward them. "Hi neighbor!" he joyfully said as he approached the newcomers. A few minutes later, the children were challenging each other to jump into the water from a small platform. "Julio, jump but without opening your mouth!" the man shouted. The child jumped into the water with his friend, and they started to play (see figure 19). The adult stayed on the creekbank observing this scene attentively. Seemingly, he tried to ensure that the boy did not swallow any drop of water. Later, we witnessed this situation several times, adults watchfully standing on the creekbank while their children swam, repeatedly reminding them to keep their mouths shut. If mouths remained shut, it seemed, everything would be fine.

Despite not smelling unpleasant odors or seeing strange colors and not experiencing adverse effects on their skin, the visitors never forget the origins of the water. On the contrary, they all assume that a certain amount of pollutants is still there, unnoticeably, capable of triggering negative consequences

FIGURE 19. Kid swimming in the Carén Creek

if they do not adopt certain precautions. Returning to the notion of inverse resistance, this constant awareness does not lead them away from the water. Rather, they continue to enjoy it by setting a threshold—their mouths—beyond which the water cannot trespass; otherwise they could end up "like the cows from upstream," as one of them told us in reference to the cows that died after the 2006 spill (discussed in the next chapter).

In taking this stance on bathing despite identifying certain undeniable risks, it is important to finally consider the people's sense of the place, the sensorial and affective components of coming to this creekbank on a summer day. As Fuentes told us, he comes to Carén Creek "because the creek makes you feel another sensation. . . . And here there are very pleasant places, so you go there, you wet your feet and you end up taking a full bath." In a region ravaged by the decade-long megadrought and the privatization of nature, to see a creek full of water surrounded by trees and vegetation was a blessing, a place to find some solace from the joint damaging consequences of neoliberalism and the Anthropocene in Chile. In contrast to empty streams and privatized or heavily polluted lakes such as Rapel, "the creek makes you feel another sensation."

Although unknown to them, enjoying that spot was also a political statement. The actors in power—corporations such as CODELCO and ENDESA, health regulations, environmental authorities—claimed that this spot, and the creek as a whole, was just a wastewater conduct, a connecting path between two processes of extraction, mineral and hydroelectric. Hence, local inhabitants should remain aloof from it, menaced by the dual risks of body pollution and legal prosecution. By swimming in the creek, they fundamentally challenged this order, turning the creek into a common, a place for the enjoyment of nature, a place for the practice of the human right to leisure.

PARASITIC LIVING

Parasitism was literally a way of life in the Carén basin. Without parasitism, if the legal demarcation of water as both private and polluted had worked, no life could have emerged on the 20 kilometers separating the dam from Rapel Lake. Well, probably some other geosymbioses would have emerged anyway, but not the ones that low-income humans could benefit from, especially in an area experiencing the most intense drought in memory. Given this scenario, farmers and visitors had no alternative to developing productive and leisurely activities in the area other than to *stay with the trouble* caused by tailings' water. It was either geosymbioses or no ecology whatsoever. Such trouble meant also life, and to keep this life farmers and visitors were willing to become intimately involved with the chemicals in the water. They deposited them on their fields, gave them to their animals to drink, passed them through their skin, and even occasionally swallowed them. All to maintain a geosymbiotic engagement with an area severely affected by the megadrought monster. Or, in other words, farmers and visitors alike have been so busy *practicing* the Anthropocene at Carén that they have had no time to think about it, much less to criticize it.

In a decentered way, through multiple human and nonhuman solidarities, farmers and visitors created some precarious spaces for simple forms of life to emerge within mining and climate disruption. In doing so, they enacted parasitism as a highly creative force, establishing unholy alliances and strange habits, from paying attention to the sores on their hands to naturalizing an industrial landscape. Instead of merely leaving/avoiding the area, the people of the Carén Creek basin invented parasitic forms of staying with the trouble of tailings in the water.

Through these practices, they also became political. This politics was not enacted through opposition, such as the anti-mining movements usually analyzed in the critical literature. In contrast, it was fought through engagement with chemicals in various ways, overcoming in the process the barriers put up by the dual extractivist politics of mining and hydroelectric waters. In line with Papadopoulos, what we found at Carén was not open resistance to pollution, but a form of politics based on "creating alternative forms of life on the ground, . . . [that] set up conditions in which movements, rather than just oppose power, installed alternative ontologies of existence."[18] These inverse resistances, then, were focused "not primarily . . . [on] direct confrontation but . . . [on] changing the conditions in which power operates."[19] Instead of protest and rejection, we find a selective cohabitation with these strange chemicals. Replacing the mainstream ethics of property and purity, farmers and visitors propose a politics of engagement with chemical beings, of becoming geosymbiotically impure, even if that engagement could end up being openly toxic, as we explore in the next chapter.

However, it is crucial not to be misled by such alternative forms of resistance. The fact that they manage to take advantage of the water in the creek somewhat does not make the violence inflicted upon them any less violent. After all, no one at CODELCO was expecting them to engage with the water in such ways. They were expected to remain silent and inert, watching passively as the narrow waterway they had used in the past was turned into a conduit for potentially toxic chemicals. As Julio Casas, one of the oldest neighbors, complained, the dam was a "pest" that had fallen upon the basin, a daily source of weakness and pain. And such a massive disruption was not temporary. As concluded by an official survey of the basin, even after the dam's closure there would be "irreversible effects" on the local environment.[20] Besides this daily pollution, a massive dam containing a tricky dragon was built over their homes, over their very lives, just waiting for a minimal mistake to unleash its fury upon them. And finally, it did.

FIVE

Life against Life

THE DRAGON UNLEASHED

"We were in Melipilla, having lunch, when his mobile rang and it was some-one from CODELCO, who says 'Horacio, where are you?' . . . it was Sunday, after Easter." In this way, Fabiola Varas, sister of the longtime caretaker of Loncha Station, started retelling the events of the Easter weekend of 2006. "Then they told him 'Horacio, depart right away and return to Carén Creek, at the Alhué bridge, because there was a mess at Loncha'. 'What happened?' Horacio asked, and I said to him 'Horacio, my god, what happened? Did the [dam] wall collapse?' I was screaming, . . . 'no, no, no, tailings went out', he replied. On the way back we stopped in an area by the creek where people had come to camp and fish for Easter and a guy said to Horacio 'hey, my friend, tell me, what the fuck is happening? Look how the water comes, is it true that the water from the mine in Rancagua is coming out?' . . . I said to him 'Horacio, my god, let's get going!', because you know, [the water] was the same as if you had emptied a truckload of cement, it was horrible."

Like Varas, everyone in the area seemed to have a story about the spill. Starting in the early hours of Sunday April 16, 2006, tailings massively flowed into Carén Creek for several hours. CODELCO talked about the release of more than 500,000 cubic meters of tailings. For many others, this number was highly conservative. The truth is that no one could say for sure how much tailings was released, mainly because no one knew exactly when the spill had started. This was true not only for people such as Varas, but also for the dam's technical personnel. Although they claimed that continual twenty-four-hour surveillance of the dam was carried out, most of those interviewed recognized that they became aware of the spill only after some neighbors from the area raised the alarm, when tailings had been pouring from the dam for several hours. To all the people involved, the spill happened below their radar. This relative invisibility was not only a matter of negligence, but also due to the characteristics of the emergency. In contrast to tailings dam breakages—which could release in a matter of minutes millions of tons of tailings—the spillage of a few hundred thousand tons of tailings into Carén Creek was rather negligible by industry standards.

Explicitly adhering to this approach, the first press release on the matter from CODELCO, published early on Monday, April 17, 2006, simply stated that "on Sunday, April 16, 2006, at the Carén dam . . . there was an environmental incident, that generated the release of an estimated volume of $0.15m^3$ per second of a mixture of water and tailings . . . into the Carén river."[1] The spill was not a disaster, not even an accident. It was a simple "environmental incident," a minor event. Although its causes were still not fully known, they were certainly discoverable through an "investigation to clarify the circumstances of this situation." Besides, its effects were rather limited, and the cleanup had already started, with the final aim of fully "restoring" the original environment of the affected area.

In developing this narrative, CODELCO was just trying to insert the spill into the conventional frame of functioning at Carén. After all, the dam has always leaked, mobilizing variable amounts of tailings downstream through the water released into the creek. Accordingly, the "incident" was framed as peak leakage, unfortunate but still within manageable parameters. The dam had been operating for more than twenty years without experiencing any problem like this, so the spill could be regarded as part of the "normal accidents" expected with large and complex infrastructure.[2]

This initial strategy rapidly failed, as most of those involved resisted seeing the spill as merely an "incident." First, there were the reactions of neighbors such as Varas, who were stunned after seeing the creekbank and some surrounding lands completely gray, "as if you have emptied a truckload of cement [onto them]." Then the relevant authorities arrived, followed by the national media, who gave ample publicity to the spill, presenting it as causing extensive pollution. The accident also coincided with an ongoing controversy about the legality of a special decree allowing CODELCO to release water with higher concentrations of certain minerals than legally allowed into Carén Creek.[3] This combination of factors made the spill a matter of public concern, whose more relevant consequence was the creation of a special investigative committee at the national congress to establish CODELCO's responsibilities for the spill.

The first matter of inquiry for this committee was causation, or establishing the exact sequence of events that led to the spill. As commonly happens in industrial disasters, establishing such causation proved to be challenging, with contrasting narratives about the process coexisting. At first, CODELCO argued that the spill had been caused by the negligence of one operative who did not realize that two water pumps, set to extract water from an area within the dam on which some works were being carried out, were discharging tailings instead of water. This well-defined and singular human mistake, which went unnoticed for four nighttime hours, had been the main cause of the spill.

This version, however, was rapidly questioned by several of those invited to speak to the committee. For example, in its first session on August 7, 2006, Raúl Leppe—then director of the regional branch of the Comisión Nacional del Medio Ambiente (CONAMA; the national commission for the environment)—presented a series of arguments showing why this version of the events was highly doubtful, from the number of water pumps involved to the total mass of spilled tailings. As a consequence, he concluded, "We have no clarity whatsoever, at least on the region, about the magnitude of the event, neither its time of occurrence. We presume that they did not coincide with the times that they [CODELCO] indicated. . . . [W]e have no data that allows us to corroborate this," a position that was largely seconded by most the members of the committee.[4]

A first counternarrative regarding the cause of the spill was presented by Federico Bustos from CICA, invited to the second session of the committee on August 23, 2006.[5] Departing from the human-caused incident initially presented by CODELCO, Bustos described the spill as "an incident that was not possible to prevent or predict." Although several risk analyses had been carried out, none included the possibility that "a wave and a resuspension of tailings inside the dam could happen. . . . I mean, the solid fraction [of tailings] escaped the dam through a wave of suspended tailings." Similar to the themes discussed in chapter 2, this wave was utterly impossible to control; it could only be endured. The dragon was back, a beast whose agency could only be partially understood.

This interpretation of the issue was cemented in the committee's next session, on October 4, 2006, when Pablo Constanzo, a high-level executive from Teniente, recognized that the disaster had happened because of "a sudden displacement of the tailings ... for a short period of time, but in a massive amount."[6] There was no human being behaving deficiently, someone who could easily embody negligence and, accordingly, be punished for that. Now tailings themselves were identified as what caused the spill. In contrast with human actions, tailings' agency appears as much murkier and difficult to understand, surrounding the whole event with a thick layer of uncertainty. As Constanzo recognized, "the explanation of this phenomenon is anything but self-evident, actually, ... [only] afterwards professional experts, when analyzing ... what had happened concluded that ... [the movement of tailings] had produced this massive spill."

As explained to us by Daniel Pérez, a member of Teniente's environmental department, instead of a normal accident, the spill "was a violent hit" whose length could not even be calculated, ranging from "maybe five hours, maybe two hours, or it could have happened in just one hour." While the failure of a human being or technical system can be studied and rapidly corrected/replaced, the dragon can only be governed through weakness, as seen in chapter 2. Several standards and devices could be introduced to help it sleep peacefully. But we can never discount its sudden awakening, as happened on that Easter Sunday, with disastrous consequences.

These consequences were especially destructive to the multiple geosymbioses that had emerged in the basin since the dam started to release water in the mid-1980s, among them the agricultural parasitisms discussed in the previous chapter. While farmers, animals, and crops had developed complex mechanisms to deal with the variable amount of tailings dissolved in the creek's water, the abrupt arrival of tailings themselves completely overwhelmed them, causing the emergence of multiple toxic geosymbioses. In the process, CODELCO suddenly realized that parasitisms were important. They had helped to keep downstream farmers quiet about the release of tailings water, especially in a time of rising environmental awareness in the country. In doing so, and probably more important, they had become a living exemplar (in a much more efficient way than the Loncha Experimental Station ever had) of the declared beneficial environmental effects of the dam, enacting a kind of mutualism whereby CODELCO could take advantage of the massive usage of Carén Creek's water to promote its operations as an example of sustainable mining. If these parasitisms were to disappear as a result of the spill, being replaced by nasty toxicities, CODELCO would have much trouble publicly demonstrating its environmental credentials.

As a consequence, the corporation rapidly developed programs to correct the situation, aiming at restoring the basin to a state previous to the spill. At the same time, one local family started a litigation process against CODELCO, aiming at obtaining financial compensation for the damages experienced. Caught between both efforts, some local ecologies paid a heavy price for humans' attempts to rule over the dragon, as we will see in the rest of this chapter.

REMEDIATING THE IRREMEDIABLE

The destructive capacities of the dragon were initially utterly disconcerting for Teniente personnel. As Contanzo admitted to the parliamentary committee, "the division . . . suffered a great identity crisis [because of the spill]."[7] This confusion was evident not only in the fact that it took several hours for the technical personnel to even notice the spill. Trying somewhat to compensate for the delay, the inaction was followed by frantic activity, especially in terms of massively introducing from the very next day heavy machinery in the creekbed to directly remove tailings. In doing so, the spill was framed as an industrial accident, needing a massive industrial cleanup operation to be controlled.

This approach rapidly backfired. First, it faced the open opposition of the people directly affected by the spill, such as Ricardo Monsalve, the head of a family that owned a large farm a few kilometers down from the dam:

> They arrived here wanting to clean up the creek, it was a Sunday, they arrived and entered [without permission], someone called me . . . and I had to come to take them out, [they acted] as if they were in their own house, [they said to me] "no! we are from CODELCO!" [I told them] "I do not care that you are from CODELCO! This is private property!" . . . In fact, when the event just happened, we allowed them to clean up . . . but with the condition that they just open four points [on our fence], enter through them to the creekbed, and remove the tailings, but, on that Sunday, they arrived not with two small machines and four trucks, there were 60 trucks, 5 bulldozers and 10 excavators, the huge ones, and [they started] razing the barbed wire, the trees, everything, then [I said to them] "no, you are going out! The agreement was that you open four points but you are razing everything," . . . they just wanted the [cleanup] to be like this, to clean our whole 2,000-meter creekbank in two days and that's all, end of the story. . . . [Because of that] I never allowed them to clean the creekbank on our property.

The little value given to the ecologies existing downstream from the dam was again visible in the brutal way in which CODELCO employees first engaged with the spill. As concluded in one of the earliest internal assessments of the environmental impact of the spill, "the predominant ecosystems on the zone are quite degraded." From such a perspective, it was easy to assume that there was nothing to protect on the creekbank, that the only task ahead was to extract tailings as fast as possible, without considering any further ecological damage that such massive intervention might cause.

Although useful to rapidly remove a significant portion of the tailings, this approach only increased the damage that the spill was inflicting on existing ecologies, especially vegetation and wildlife. In line with Monsalve's comments about uprooted trees, Pedro López, the local representative of the public Servicio Agrícola y Ganadero (Agriculture and Livestock Office, SAG), who visited the site a few days later, wrote in a report that "the first thing that caught our attention was the mortality of fish, mainly carp and silversides, and the affection of nesting sites of endemic birds, . . . [such as] taguas, little egrets, cuca heron, huairavos, yecos, among others."[8] These impacts were the result not only of the spilled tailings but also of the "drying up of the river for cleaning," meaning the abrupt disappearance of most water in it. From the standpoint of an industrial cleanup operation, these beings, which had developed quite sophisticated ways to engage with tailings water, were suddenly deemed irrelevant, being readily sacrificed in the search for chemical purity.

When the spill became a matter of public controversy, however, it was not presented as an industrial accident. On the contrary, for the media it was undoubtedly an "ecological disaster," usually involving commentaries and pictures about the damage done to local ecologies, both agriculture and wildlife. As a consequence, CODELCO was forced to change its initial brutal approach. As an executive declared to the parliamentary committee, the spill "has been a situation that has afflicted us and regarding which our reaction has been to remediate the damaged caused."[9] This was no longer an industrial cleanup operation; it was a "remediation," a term that carries a completely different sense.

In its most common usage, *ecological restoration* is the sustained effort "to bring back particular landscape layers, to remove ecological damage and disturbance and rebuild ecological communities, returning them to a healthier state."[10] In practice, this effort has frequently translated into the explicit aim of bringing "landscapes back to something approaching their structure and functioning prior to human influence."[11] Ecological restoration usually starts by setting some baseline conditions, normally understood as the environmental conditions existing in an area before human contact. Afterward, several corrective measures are introduced to try to align the current state of the environment with the baseline.

Regarding Carén, the application of this conventional notion of remediation immediately appears problematic. As shown in previous chapters, there is nothing conventionally "natural" about Carén. Since tailings water started to flow from the dam in 1987, local environments have been transformed beyond recognition. And probably more importantly, the transformation has not followed the usual path of human-induced degradation. On the contrary, the dam and its water have caused the emergence of lively geosymbioses all over the basin, even including "nature" hotspots such as Laguna de los Patos, a lagoon that emerged 1 kilometer downstream of the dam, becoming in the process a feeding and nesting place for several kinds of wild birds. How can we "remove ecological damage and disturbance" in ecologies where human-induced "disturbance" has been commonly the direct opposite to "damage"?[12]

Given these antecedents, restoration at Carén was not going to be based on conventional notions about the existence of a "balance of nature" prior to the spill.[13] The basin was largely understood as being heavily transformed by human activity, and the main aim of remediation was to restore it to a previous status, in which human interference was assumed to have enacted a happy coexistence, as discussed in chapter 3. By removing the need to set up a proper "natural" baseline for the basin, this approach gave prominence to two interrelated thorny issues: (1) what exactly to restore and (2) when such restoration would be considered complete.

The definition of what to restore was directly dependent on the perception of what had been damaged by the spill, an assessment that varied with the passage of time and the actors involved. At the beginning, as seen earlier, professionals such as López from SAG put their focus on the damage itself, on the dead fish and displaced birds and the images of the former "natural reserve" Laguna de los Patos, now fully covered in tailings. Given the scale of the damage, the remediation effort would be profound, involving important costs and extensive timelines.

To provide another narrative, CODELCO hired Fundación Chile, a prestigious technical consultancy institution, with the stated aim of "objectivizing the qualitative assessment of the spill's impacts."[14] Through this "objectivizing," which included carrying out a census of the local flora and fauna, a more limited picture of the spill's impact emerged. Regarding animals, the first report concluded that "impacts on water birds cannot be attributed to the incident. No bird mortality was observed and it is known that they have moved to other areas. Dead carp were observed . . . [possibly] related to a drop in water level."[15] Regarding vegetation, it was argued that "the impact . . . is of little significance in the quality of the ecosystem, as well as in species and in number of affected specimens." Every possible damage that appeared as menacingly diffuse on the SAG report was given a tranquilizing diagnosis in this second assessment. Instead of guesses, this was a census carried out by a highly regarded technical institution, which disregarded most major environmental damage besides some dead carp (fish that, as seen in chapter 2, were utterly valueless anyway) and some cattle. Although such reports could be seen as merely "special interest science," they were taken by the corporation as the ultimate word on the matter, framing the spill's effects as "temporal and reversible, after the cleanup process," as Constanzo told the parliamentary commission.[16]

Based on this assessment, Fundación Chile developed a detailed remediation plan aimed at eliminating "the alteration produced in the natural and human environment as a consequence of the accident."[17] In order to be able to assess this alteration and evaluate its remediation, they adapted a well-known analytic scheme developed by Leopold and colleagues to evaluate the environmental impact of human-induced alterations.[18] This scheme divides environmental impact into five well-defined "attributes," assigning to each one a limited number of parameters, from the "degradation of vegetal communities" to the "alterations of traditional ways of living."[19] In practice, this matrix was translated into a series of tables, on which each parameter is evaluated in terms of a qualitative scale ranging from high/extensive/permanent to low/punctual/temporary impacts.

In contrast with the grizzly images of destruction and death portrayed by the media and the SAG report, these tables offered a neat take on the environmental impacts of the spill, a synoptic view of all the different aspects clearly divided into categories, attaching a single qualification to each one. Regarding these aspects, a single *hito de cierre* (closure milestone) was established, usually involving a relative increase in the valuation scale, after which the issue was treated as utterly solved. Through this scheme, Fundación Chile offered order and a clear path forward, a step-by-step method toward a fully remediated (and normalized) Carén basin.

Although quite reassuring for CODELCO, this neat scheme worked quite poorly for local ecologies. As seen in the previous chapters, local geosymbioses were anything but neat. Starting from the carps and algae within the dam, life in the basin emerges out of the entanglements between entities that should not entangle, especially living beings and the chemicals forming tailings.

However, these geosymbioses, due to their inherently transboundary nature, actively resisted being located within neat categories such as the ones developed by Fundación Chile. Given the low value attributed to the organisms from the start, it is not surprising that their resistance was turned into ultimate invisibility. Especially regarding wildlife, given that "few specimens were observed, due to the anthropic pressure to which they are exposed," any further negative effect the spill could have had was deemed irrelevant.[20] The basin was heavily impacted before the spill anyway. The only two species regarding which certain damage was acknowledged were carps and African frogs (Xenopus laevis), but it was immediately noted that "both species are introduced," meaning that they are "invasive species" in the usual parlance of ecology, so no measure was needed to repair the damage. Actually, from the usual point of view of conservation biology, such damage could even be seen as a good thing, allowing this ecosystem to get rid of some of these undesirable invaders.

Given these framings and exclusions, it is not surprising that only a few months after the spill, in October 2006, Fundación Chile published a final report on the process, affirming that "the measures taken by the company . . . to fix the situation and reduce the negative impacts and environmental risks . . . have been correct. Finally, it is estimated that the environmental impact caused by the accidental tailings spill is reversible in nature, as long as the suggested rehabilitation measures are effectively completed."[21] In order to materialize this claim, the report included a series of before-after sequences of pictures, vividly showing how areas in which the greyness of tailings was ubiquitous in May now appeared as fully recovered, even including a return of creekside vegetation.

Following from this, the full remediation of the basin and the restoring of CODELCO's environmental credentials were deemed to be substantially finished by the end of 2006. As affirmed in the institutional yearbook for that year, "Currently, the affected area shows practically the same condition it had before the spill, which testifies the reversibility of the effects of the incident, as well as the promptness and effectiveness of the mitigation tasks."[22] By the end of the year, everything seemed to be back to normal; the spill was merely an "incident," little more than a precautionary tale. Then the disgusting image of a dead cow with its stomach full of tailings brought it all back to life.

On October 12, 2017, the Chilean Supreme Court issued its final verdict in a lawsuit that the Monsalve family had brought against CODELCO after the 2006 spill, finalizing a legal process that had lasted more than ten years. The verdict affirmed that taking into account the death of animals, the loss of vegetation, and the loss of crops, among other things, there was no doubt that the Monsalves experienced damage as a result of the spill. Given this, the Supreme Court ordered that CODELCO must compensate the plaintiffs, definitively confirming the position previously taken by lower courts.

As seen earlier, the Monsalves had been utterly enraged about the spill from the very beginning, not only because of the ecological damage but also due to the utter disdain with which CODELCO had treated them and their fields in the cleanup process. Given their relatively better financial status— they owned a 614-hectare farm, mostly devoted to producing vegetables and, especially, raising cattle—they were able to resist the attempt by CODELCO to turn the damage they had experienced into just another CSR scheme.[23] They even denied CODELCO entrance to their property to remove the tailings accumulated on the creekbank in front of their property, until they were forced to do so by a local court.

As a consequence of the extended presence of tailings on the creekbank, and the fact that CODELCO stopped releasing water from the dam during the cleanup, the Monsalves' cows started to drink directly from puddles formed over the spilled tailings. The consequences of this practice were rapidly evident; as Ricardo Monsalve told us, "after a week a calf died, two days later, a second cow, then every day one more, some days it was three, four, five dead cows.... [I]n the first year more than 100 died." When some lawyers appeared on their doorstep a few weeks later, offering to represent them in a lawsuit against CODELCO for damage derived from the spill ("they came here and offered themselves," in Ricardo's words), they duly agreed. The lawsuit was just one more component of a sustained campaign of protest they had carried out since the spill to seek compensation, and probably the one from which they expected the least.[24]

As recognized in the literature, culpability in pollution-related lawsuits, known as "toxic torts," is extremely difficult to prove.[25] This difficulty is derived from three interrelated factors. First, even in general terms, "understanding the properties of . . . [toxic] substances and assessing any risks they pose, requires even more subtle scientific expertise and studies than for other areas of inquiry. And they usually must be conducted on the frontiers of existing scientific knowledge."[26] Second, exposure to such toxicants "is notoriously difficult to characterize with precision, especially in cases involving environmental toxicants."[27] Finally, even if exposure is proven, to show in a convincing way that actual harm has occurred as a direct consequence of such exposure is extremely difficult. Usually "it can take years to have clues that substances cause harm, and even longer to document the cause of damage."[28] All this without even considering the politics of toxicology, a science that has been historically oriented toward facilitating environmental pollution rather than controlling it.[29] Even in the few cases in which the evidence is rather solid and comprehensive, trials end up concluding that "perpetrators of environmental harm . . . [were] reckless and/or negligent rather than malicious."[30] As a result, many (if not most) toxic torts end up without offering any kind of redress to the victims.

In the Chilean case, the situation is not different. Although the national constitution includes an article that "warrants to all people the right to live in an environment free from pollution" (Article 19, number 8), in practice, this mandate has not resulted in higher rates of success for toxic torts. This is so because the article, although "allow[ing] the penal protection, does not suppose an absolute mandate for penalization, and this penalization, in case of being granted, could be limited . . . by the need to recognize other rights."[31] Given that the constitution was promulgated during the dictatorship, at a time of high neoliberal fever, in most cases rights regarding private property and entrepreneurship tend to prevail over environmental rights in courts.[32] The fact that this case was presented by the members of a single family against a large and powerful public corporation such as CODELCO makes its success even more remarkable.

What explains this unlikely success? Following the threefold model for toxic torts, we could start by recognizing that the plaintiffs found in tailings a quite useful toxic "villain." As seen in chapter 1, since the start of large-scale industrial mining in Chile at the beginning of the twentieth century, the collapse of tailings ponds has been the cause of some of the worst industrial accidents in the country, the last of which happened in 2010.[33] There is also already a relatively large body of research regarding the toxic capabilities of tailings, especially related to human and animal health.[34] A second advantage for the plaintiffs was that the usually complex issue of showing the existence of a pathway between the pollutant and the affected entities was rather straightforward in this case. The tailings found on the Monsalve's property could not come from any other place than the Carén dam. Besides, several technical reports and the parliamentary inquiry concluded that the release of such tailings was culpable negligence on the part of the corporation. Then, the presence of a toxicant and its pathway toward the affected entities were rather easy to prove. However, the more complex component of this model was still pending: harm.

In the initial lawsuit, presented to the Court of First Instance of Melipilla in July 2007, it was argued that the spill produced two kinds of harm for the plaintiffs. First was an emergent harm (*daño emergente*), referring to the direct damage caused by the spill. Second was the loss of profit (*lucro cesante*), or the economic losses experienced due to the potential profits that the affected would not be able to get due to emergent damage. Both damages were argued in the following way:

> CODELCO's tailings spill . . . has caused enormous damages and economic losses to [the plaintiffs] . . . such as damages caused by contaminating with tailings existing soils and crops, which were damaged and rendered useless by the tailings, . . . and by the death of numerous animals resulting from the contamination of the waters in Carén Creek that were used for animal consumption and due to the weight loss of those who survived and the deterioration of their physical and sanitary conditions. . . . Animals killed after the spill add up, to the date of filing this lawsuit, to 34; there are also about 400 Hereford cattle that have not been sold by their owners, due to the deterioration caused by the aforementioned tailings spill in their physical and sanitary conditions. . . . The Servicio Agrícola y Ganadero [SAG] . . . has received and verified the complaints from the affected regarding the death of these animals.[35]

From the very beginning, the dead cattle were the main means by which the plaintiffs argued for the undeniable harm they had experienced and the rightfulness of asking for financial compensation.[36] In making this argument, the lawsuit understandably left out any reference to the beneficial geosymbioses in place at Carén for decades, especially the fact that the cows had been exclusively (and unlawfully, as seen in the past chapter) drinking water containing variable amounts of tailings their entire lives. Instead, the complaint focuses solely on the toxic geosymbioses that emerged right after the spill and ended up causing the deaths of multiple cows. The spill appears as clearly disrupting the "balance of nature" in place in the basin, causing multiple damages to the biological entities, especially cattle. Their deaths and the "deterioration of their physical and sanitary conditions" were used throughout the trial to materialize the damage done by CODELCO's mistakes.[37]

However, as the last part of the preceding extract reveals, this embodiment of damage did not mean that the dead cattle were recognized as the party harmed by the spill, the "victims" of CODELCO's misbehavior. As the quote duly explains, here the ones suffering were not the cows, but their owners, who saw their current (and future) profits reduced as a result of the spill. The cows, on the other hand, were merely damaged property. In line with the explicit anthropocentrism of most legal bodies worldwide, a tendency to which Chile is no exception, the cows were not seen as subjects of the law, but merely as objects whose damage affected some other (human) subject.[38] The dead and "deteriorated" cows were merely an indicator, a way for tailings to become toxic, the source of damage to humans that must be compensated. In practice, "the animal seems to designate no more than a kind of occasion for a harm recognized by law."[39]

In order to support this position, the file included several institutional and scientific documents, such as the necropsies done on some of the dead cows by the prestigious Universidad de Chile or a technical report from the department of environment health of the Ministry of Health. While the first document talks about the "suggestion of toxic damage" to the analyzed corpses, the second claims that "the immediate effect [of the spill] has been the death of some animals that water on the river."[40] To give a more graphic weight to such claims, the file also appended more than seventy images of the dead cows, some of them even showing the corpses during the necropsies. One of the most vivid ones depicted the open stomach of a cow full of tailings (see figure 20).

The technical reports and images enacted the dead cows as vivid exemplars of the damage done by the spill to the Monsalves' property. Other than that, little was heard about them. There are only a couple of references to the more than two hundred cows experiencing suffering and pain due to the toxic geosymbioses happening inside their bodies, their utter misery and agony before ultimately dying. These cows are silent and immobile, no different than tailings themselves. The only version of the cows that mattered here was the one that presented them as damaged property; any other attribute was invisibilized, considered irrelevant by the plaintiffs and their lawyers.

As could be expected, CODELCO's defense refuted most points posed by the plaintiffs. Their starting point was that, after remediation, pollution no longer existed in the area. Understanding pollution and remediation as opposites, the fact that remediation was completed and certified by Fundación Chile appears as enough proof that it was incorrect to still speak about

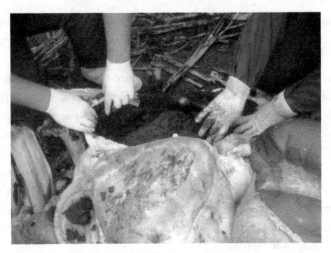

FIGURE 20. Open cow stomach during a necropsy

pollution on the Carén Creek. This position was reinforced by statements made by executives from Fundación Chile, according to which "the concentrations of [chemical] elements of relevance, both in the water and in the sediments, returned to levels existing prior to the incident."[41] However, no information was ever provided about how much these "prior to the incident" concentrations were, probably to avoid entering into the murky issue of CODELCO's historical release of industrial water into the creek.

Moving into attack mode, CODELCO's defense argued that if pollution in the area continued it was only as a consequence of the reckless behavior on the part of the plaintiffs, given that they refused from the very beginning to allow CODELCO personnel to enter their property to carry out remediation, an opposition that "allows to suspect about the causes and scope of the damages denounced by the plaintiffs."[42] These suspicions were especially focused on the issue of the dead animals, as noted by a declaration of the CODELCO employee in charge of overseeing the cleanup: "The damages [to animals] that may have occurred took place on the two or three days after the event, . . . when there was a refusal on the part of the [Monsalve] brothers to give them [clean] water, at the beginning they opposed. . . . the rest of the farmers were given [clean] water . . . for their cattle, and with them there was no problem."[43]

Expanding on this line of argumentation, CODELCO's lawyer suggested that tailings intake was not the real cause of death of the animals. In this regard, he quoted a statement provided by one of the witnesses enlisted by CODELCO, affirming that "CODELCO had a veterinarian who monitored the health condition of [animals on] the entire area . . . [including] the plaintiffs' property, *who refused . . . to allow a study of their livestock as well as to carry out examinations of their dead animals.*"[44] Using a well-known defense strategy developed in tort law regarding tobacco consumption, CODELCO was trying to present the Monsalves as victims *but* irresponsible, greatly weakening their compensation claim.[45] CODELCO was not denying that the plaintiffs suffered certain damage caused by the spill, especially materialized in the four cows that died early on. But beyond that, all other damage was caused by their own irresponsibility in not allowing CODELCO personnel to clean up and remediate their land. Given how difficult it is to prove culpability in a toxic tort, "a test case plaintiff is not an individual who has voluntarily courted risk."[46] And, as CODELCO was trying to show, this was clearly the case here, as the plaintiffs knowingly courted risk by refusing CODELCO's water and remediation.

Against most expectations, including the plaintiffs', on June 21, 2016, the local court of Melipilla ruled against CODELCO, demanding it pay compensation to the Monsalve family, a judgment that was confirmed by the Appeals Court nine months later. Besides concluding that CODELCO was culpable for the spill and the general damage to the environment it caused, the ruling concluded that

> . . . the water of the Carén and Alhué streams was necessarily contaminated, thereby affecting the flora and fauna of the sector, as well as livestock and agriculture productive processes on riverside property. . . . Of the damages produced by this incident, they account for the death of animals, cattle, horses, wild animals, fish, bird migration, wildlife; in turn there is damage to the flora along the riverbank. . . .
>
> Given these antecedents, there is no doubt that there are damages on the plaintiffs' premises, as well as on their flora, livestock and crops due to the environmental event under consideration which was the defendant's responsibility; in fact, there are deceased cattle, affected crops, flora adjacent to the leased channel as indirect damage due to the cleanliness of the channel, etc. . . . the evidence of both parties indicates the existence of damages regarding not only to the general environment affected by the illicit, but with respect to the precise assets of the actors.[47]

Agreeing with the narrative presented by the plaintiffs, no recognition was given to the previous presence of chemicals from tailings in the local water.

Through such purification, the spill was enacted as an utterly contaminating event, importantly affecting local ecologies, both wildlife and agricultural.

In taking this position, however, the court was not motivated solely by the environmental damage done by CODELCO. Although such damage was recognized, the matter at hand was not a ruling or compensation directed toward a proper remediation of the affected environment. The issue was how to financially compensate for the damage to the "precise assets" of the plaintiffs, again considering the almost two hundred dead cows as a matter of lost private property, not regarding animal suffering or anything related. In taking this position, CODELCO's argument regarding the plaintiffs' responsibility for their deaths was not just defeated; it was not even mentioned by any of the three courts overseeing the case. Legally, the dead cattle were not treated as sentient beings who experienced pain and suffering due to some human actions, but merely as damaged goods. Given this, the final matter was not how to compensate for the harm done to those animals, but how to establish their value as lost property.

To deal with this latter issue, the plaintiffs submitted further documentation stating the compensation that they should receive. Specifically, they proposed that CODELCO pay them CLP 3,500 million (around US$4.38 million) because of the damage produced by the spill. Calculating the emergent harm was relatively straightforward, amounting mostly to the direct calculation of the market value of the lost cattle and crops. Quantifying lost profits, by far the largest portion of the claimed compensation, was much more challenging. This complexity derived not only from the inherent difficulties of valuating nonhuman entities, but also from the fact that the value needed to be actualized, a process in which several noneconomic assumptions were introduced.

To carry out this task, the plaintiffs hired an economist to make a report on the matter.[48] As the main mechanism to calculate the loss of profit, the report declares:

> To estimate the loss due to loss of profits per year [since the year 2006], the existence and future temporary projection of a group composed of 543 cattle must be considered, ... the annual flow generated by the production of kilos of weight must be determined (meat) from such a group, denominated as the annual Animal Weight Gain (GPA [in Spanish]) of the group, sent to the Fair in Melipilla, which sets the reference price for such a product.... This base group allows the generation of 211 calves, with their respective GPA, as well as the GPA of steers and fat cows that are 324 cattle (275 + 49)....

The animal weight gain of a Pregnant Cow corresponds to a calf whose birth weight (PAN) is 50 kilos after nine months of gestation and also 20 kilos in the first three months, daily GPA of 0.22 kilos per day, this implies that a pregnant cow generates 70 kilos of weight gain per year.[49]

Detaching the issue even further from their suffering, the dead cows are now simply enacted as kilos of meat, meat to be sold on the market. The price of this meat for the year 2006 (updated to 2018 prices) generated the ultimate valuation of the lost profits, a full quantification of this damaged ecology in which no extra-economic consideration toward the living beings that were impacted, even killed, by the spill is ever introduced.

WHAT KILLED CARPS AND COWS?

At the time when our fieldwork was carried out, there were no visible reminders of the spill. Upon visiting the area, it was impossible to notice any sign of the dragon and its fury. Most neighbors still remember it, but as a fading bad memory, usually replaced by other, more urgent concerns. For CODELCO personnel, the spill has become mostly an exemplar, a moral story about the terrible consequences of being sloppy in their daily management of tailings and the motivation for a complete overhaul of their waste management practices. What is conspicuously absent from the consideration of all those involved are local ecologies, especially the deadly encounter between the dragon, multiple carps, and more than two hundred cows.

This invisibility is not accidental, but was embedded into the two main approaches devised to deal with the spill, remediation and compensation. In the first approach, invisibility resulted from a demand for ecological purity, for neatness and nativism, to be taken into account, a move that automatically made invisible the messy geosymbiotic entanglements characterizing this basin. In the second approach, invisibility emerged as the denial of personhood to the dead cows, seeing them merely as damaged property, as a few kilos less of meat to be sold in the market. Unable to properly *see* these ecologies, remediation and compensation were enacted as mostly a normalization of the human entanglements disrupted by the spill, little else. All the masses of living others, biological and geological, were simply left out of the arrangement, the denial of a voice rendering them just inert *things*, out there, to be appropriated or discarded.

Such an outcome poses a larger question: What value does geosymbioses really have in residualism? Far from narratives about sustainability and happy coexistence, this chapter teaches us that ultimately geosymbioses are highly precarious, being easily regarded as something that does not have any value in itself, merely a side effect, at most a weird "externality" that can be sacrificed without much remorse in the pursuit of more important ends, usually monetary profit. And this misevaluation applies not only to big corporations, the usual suspects, but also to less powerful actors, such as the Monsalves. As seen throughout the book, geosymbioses appear as occasions for the flourishing of life in the most unlikely of circumstances, offering paths to start inhabiting the inescapable ecological ruination surrounding us. But even with their considerable resilience, the long-term survival of those involved in geosymbioses is far from secure. In most ecologies worldwide, geosymbioses are highly threatened by worldviews that value all things, regardless of their specific nature, for their purity, either ecological nativism or clear economic value.

So, how do we care for geosymbioses? Or, more specifically, how do we regulate their interactions, especially when they generate friction and damage? When dragons attack carps and cows? In the search for more fruitful ways to engage with geosymbioses, to nurture their healing powers *and* be attuned to their destructive ones, we certainly need to move away from ready-made intervention packages, such as remediation and compensation. In doing so, however, we cannot just hope for such monsters to magically disappear, aiming to return to some imaginary idyllic past of "natural" equilibrium. There is simply no cabin in the woods for humanity to take shelter in. We (still) have some woods, for sure. But we do not have a cabin. We are out in the open, with the trees and our monsters, and our chances for survival rest on learning how to live with them.

Symbiopower

In January 2021 we were back at Carén, for the first time in years. Our aim was to check whether the situation there continued to be as described in this book, with tailings water providing the basis for the proliferation of minerals and (sometimes) biological entities we understand as geosymbiosis. However, upon arriving at our first point of contact with the creek in Valdebenito—one of the places where we had engaged with the holidaymakers seen in chapter 4—the situation has changed radically. Instead of a body of water in which people happily swam, the creek was reduced to a couple of large puddles. Obviously there was no sign of the happy swimmers of yesteryear. The whole place gave an impression of complete abandonment, only the remaining litter signaling its past usage as an informal summer resort. Although the puddles suggested that some water had flown there not long before, to find the stream missing was a shock to us.

The situation was similar some kilometers up the road, near the Valdebenito camping area. Again, the image was very different from what we had seen before: where there used to be a narrow but constant watercourse, there was now an empty creekbed. The image was not of utter desolation; although the creek was devoid of running water, the creekbank was covered with weeds and plants, suggesting that some water still remained there. But there was no flowing water. Maybe we had been wrong all along? Maybe the different geosymbioses we have described in these chapters were a thing of the past? Maybe extraction has finally become extractivismo at Carén, merely a source of harm and pollution? We continued approaching the Carén complex, mulling over these somber thoughts.

To our utter relief, the situation suddenly changed a few kilometers up the valley. At the first bridge just after the turnoff toward the complex, we saw that the creek was carrying water in an amount quite similar to what we remembered. Surrounded by greenery, the water looked clean, even transparent. Approaching it, we could even see plenty of small fish swimming in it. Above the sound of flowing water, we could hear the customary noise of the farmers' pumps extracting water to irrigate their crops. We relaxed somewhat. Here the creek flowed as ever, still producing on its way multiple geosymbioses.

However, the missing water downstream troubled us. We could not identify the cause of this phenomenon or whether it was permanent. From the puddles and the vegetation on the creekbank we hypothesized that it was only a temporary situation, that probably the water would flow through Valdebenito and downstream soon enough, reactivating the geosymbioses existing there. But the situation could just as well be otherwise. Maybe what we witnessed that day was the beginning of the end of such geosymbioses, even toxic ones, with the basin slowly drying up as a result of the lack of water, the megadrought finally catching up with Carén.

This is, after all, the planned future for the basin. As stated in the Mine Closure Plan for Teniente, after the mine closes in 2064, "due to evaporation and the fact that tailings will no longer flow, the volume of water in the lagoon will be reduced," disappearing in due course.[1] The end of flowing tailings would certainly have devastating effects on most living beings in the basin. Given the reasonable expectation that in a future context of fairly advanced climate change the megadrought would only have become more intense, to stop discharging tailings at the dam would not only mean the disappearance of the lagoon but more importantly of Carén Creek itself. That would finish most of the geosymbioses explored on this book, from the carp in the lagoon to the farmers daily extracting water from the creek to irrigate their crops.

Other geosymbioses would remain, only becoming intensified in some cases, such as the toxic entanglements of minerals and bacteria that produce the acid mine drainage that would massively pollute local soils and aquifers. Or the terrifying agencies of a dragon that, only subjected to irregular surveillance once the mine has closed down, will be able to roam freely downstream from the dam, helped by the fact that central Chile is one of the most seismically active regions on Earth. We could easily imagine that some decades after Teniente's closure, the Carén basin will be similar to a massive land art project, with wave after wave of abandoned tailings flowing down from the dam, covering abandoned agricultural fields and the empty creekbank, only here and there forming nested spaces for some proliferation of minerals and biological entities, for further geosymbioses.[2]

The picture of Carén that we got from this speculation about its after-closure future differs little from any other ruin of late industrialism, a place beset by pollution and death. As a consequence, CODELCO becomes just the usually exploitative large corporation leaving its devastation behind, and local communities, human and nonhuman, finally enact the role of sufferers usually assigned to them in extractivismo stories. As soon as tailings stop flowing into the dam, Carén's exceptionalism vanishes.

Avoiding this outcome will require much more than further CSR schemes or open condemnation. It would require fundamentally changing our understanding of anthropogenic compounds such as tailings and the ways we engage with them. A first step should be to understand mining residualism—or the logic that makes invisible materially and as a matter of concern the massive amount of residues that industrial mining entails—as resting on two failed premises. First is the notion that tailings easily sediment into geological strata. As we have seen, tailings are continually moving, leaking, and percolating, establishing all kinds of unholy alliances along the way. Second is the belief that tailings could easily coexist with biological entities, especially agricultural ones. Such coexistence exists—what we have called geosymbiosis—but is far from easy and mutually beneficial.

When these premises are discarded, residualism finally appears as what it is: a technopolitical logic in which the manufacture of shiny new mineral-based devices and infrastructures for some (usually affluent, Western, white) is only a side effect of the production of mammoth amounts of mineral residues, residues that end up impacting multiple—already vulnerable—beings and ecologies. This is true not only for human communities living near places such as Carén, but also for the multispecies ecologies they belong to, reaching well into the depths of the earth. Given the scale of this assault, geosymbioses like those seen in this book emerge as a last resort for many beings, as the only way—usually a toxic one—of continuing to exist in places of capitalistic devastation.

To see residualism as what it is should lead us to also start facing the ugly truths of the mineral modernities that have guided the development of Western societies since the Industrial Revolution. Even the shiny new "sustainable" modernities of electric cars and renewable energy ultimately rest on the extensive extraction of different kinds of minerals, producing massive amounts of residues that accumulate in thousands of places like Carén all

over the world. This is the ultimate "infernal alternative" of current capitalism, that any "transition" toward the future promised by the sustainability
program necessarily rests on wrecking lives and worlds throughout the globe,
frequently forcing them into toxic entanglements just to keep surviving for
a while.[3] As long as we use any product including copper, we are part of the
conditions that generated Carén.

A possible escape from residualism and copper modernities, however,
cannot be the search for ultimate purity, in the form of either a return to
an ideal past or a complete ecomodernist restoration. Regarding the first
option, environmental violence at Carén did not start with the dam but can
be traced back (at least) to when the Spanish conquistadores usurped the
lands of the resident indigenous communities in the eighteenth century,
an exclusion cemented by the hacienda system at play until the arrival of
CODELCO in the mid-1980s. A past of supposed equilibrium in the area is
long gone, if it ever existed. Regarding the second alternative, given the scale
of the disturbance caused by the millions of tons of tailings piled up inside
the dam, no proper restoration appears feasible, even in the unlikely case that
the funds were available. Even sophisticated forms of remediation—such as
phytoremediation—would deal only with the surface of the dragon, leaving
untouched the rest of its massive body. Besides, as with any other industrial
endeavor, an ecomodernist restoration at Carén would produce new residues,
to be accumulated somewhere else, new sources of toxicity and damage. So
any intervention in a place such as Carén should go more in the direction of
maintaining certain existing conditions and actively fighting others—such as
toxic geosymbioses—more than any proper remediation.[4]

Instead of seeing it as backward or impure, we should approach Carén as
a place where our probable future is already being played out. Against usual
Western notions that the disaster of environmental collapse is (always) in
the near future, in places such as Carén it happened decades ago.[5] And this
disaster is not related solely to the arrival of tailings, but goes back in time
to the colonization of these lands and upward to a changing climate and the
monstrous megadrought.

But several forms of life have persisted, even flourished, within this devastation. Given this setup, Carén could even be seen as a place in which a
precarious version of a "good" Anthropocene has already emerged.[6] It is not
a very appealing one, especially based on conventional notions about what a
healthy or beautiful environment is. Neither is it very fair, as the usual cast
of undesirables is still carrying most of the toxic burdens. But at least it has

allowed multiple beings to maintain several critical ecological functions in the face of massive anthropogenic disturbance. Similar to current analyses of resilient indigenous worlds, and adapting freely from Viveiros de Castro, at places such as Carén "we will have a lot to learn from … [entities] who, nonetheless, have managed to abide, and learned to live in a world which is no longer their world 'as they knew it.'"[7]

What does Carén tells us about our future? First, once we live in an environment saturated by anthropogenic chemical compounds, we become entangled with them. As seen throughout this book, tailings are entities endowed with a never-ending drive to proliferate, to expand, establishing all kinds of unholy alliances along the way. As a result, through decades of intimate (but problematic) coexistence, tailings at Carén have become "significant others" for most of the entities populating the valley: human, animal, vegetal, mineral.[8] Through such entanglements, they have not only achieved their aim to keep proliferating; they have allowed some of the biological entities with whom they have engaged to keep existing, even to thrive in an environment that pushes them to leave or die. Tailings cause damage, no doubt about that. And such damage sometimes leads to death. But they are also vehicles for desire, for the desire to continue living in a heavily intervened environment. Tailings negatively impact many lifeworlds, but at the same time they open the door for the proliferation of other worlds, worlds of gray and green. We might not like many of the worlds tailings help create, but they emerge anyway.

In proliferating in such a way, especially against the wishes of powerful corporations such as CODELCO, tailings should be seen as exercising a capacity that we can call *symbiopower*. This power derives from the relentless urge toward proliferation and geosymbiotic entanglement crisscrossing ample varieties of encounters between living and nonliving beings, a drive to relate and interchange, to become together. Signaling matter's desire for promiscuity, a mutual lust that constantly redefines identities, symbiopower is at the very center of the earth's vitalism, from the microbial to the global.

A distinctive feature of such power, which differentiates it from biopower and geontopower, is that human beings are not in control, most of the time.[9] Although usually related to anthropogenic interference, symbiopower is not a power that human beings could fully adopt and use. It could become intermingled with forms of bio and geontopower, for sure, but at its heart it is characterized by an utter disdain, a radical indifference, toward human ends. As we have seen throughout this book, the geosymbioses seen in Carén had little respect for most kinds of human governance or management, frequently making a joke of containment measures and techniques. Irrespective of human intentions, tailings' geosymbioses have their own aim—to proliferate through entanglement—and they are going to pursue it using every means available, whether human beings like it or not. They are going to pursue their goal, even if by doing so they end up killing some of the biological partners—including human beings—through which such proliferation was originally achieved.

As a consequence, geosymbioses will never fully comply with conventional notions of management or governance.[10] After all, the "monster always escapes."[11] There is always a remainder, something that has not been (and could not be) taken into consideration. Symbiopower can sometimes be harnessed to achieve human ends, such as when bacteria are used to refine minerals. But in many other situations, symbiopower can go directly against human desires, as happened in the case of the spilled dragon or when crops resisted calls to enact a happy coexistence with tailings water. In such cases, human agency is usually centered not on governing *through* symbiopower, but on the demanding task of adapting to its consequences. Any form of human project becomes, most of the time, the careful and (always) partial attuning between human ends and a plethora of symbiopowers that humans cannot fully understand, much less control.

Something similar could be said about the belief regarding symbioses' morality. As seen throughout the book, there is nothing inherently good in geosymbioses. Certainly many geosymbioses are mutualistic, providing benefits to both biological and chemical partners. But this is not compulsory, and not even the most likely consequence. Quite frequently geosymbiotic entanglements are characterized not by cooperation, but by plain coercion. And such entanglements can be quite stable over time, even if in the process they cause sustained damage to some partners, leading them slowly to death.

More than a further turn in the modernist aim of human control over the nonhuman world, human relationships with geosymbioses should be framed as a *politics of weakness*, or a politics whose strength (quite paradoxically) rests on the recognition that not all the relevant actors can be accounted for or, even if they are, can be dominated or even inspired by human desires.[12] Against optimistic arguments for the establishment of a novel "natural contract" with the nonhuman world, most of the time not even a more pragmatic "parliament of things" can be established with the entities entangled in a geosymbiosis, mostly because such entities lack any kind of "language" that we could understand.[13] Even if they cause us extreme pain and misery, there is no way we can meaningfully communicate with most of these entities, especially at the microbial level. Geosymbionts have geosymbiosis, not politics. Especially when the Anthropocene faces us with massive geosymbiotic processes that menace to wipe vast amounts of human life from the face of the earth, a politics of weakness should be mostly seen as the search for tentative paths for human survival in the face of these uncontrollable forces.

As a rather pessimist program, we can think of such politics as aiming for the development of different forms of *careful parasitism*. Like farmers and holidaymakers in Carén, we must learn to parasite carefully, always watching for when our parasitism is causing too much damage to the beings upon which we depend and/or annoying some powerful monster that might want to destroy us. We may weaken, but not destroy, our multiple nonhuman hosts on Earth.

In learning to parasite carefully we should avoid the trap of attaching moral value to entanglement, naively assuming that the problems geosymbioses pose for human and nonhuman survival and well-being will be automatically solved by their inner morality or goodness. Geosymbioses do not care about human morals or well-being; they only care about proliferating. And this proliferation, as seen in several chapters of this book, could well lead to violence and suffering, even ultimate death, for some of the participants. So commonly, a politics of weakness should also put its focus on devising ways to disentangle, to escape from damaging geosymbioses in the least painful way, especially those that damage individuals and groups already affected by other unequal entanglements, both biological and political.

Given anthropogenic compounds' multiple threats to human and non-human well-being, a politics of weakness should start from the very basic question of whether to produce them at all, especially entities as massive and promiscuous as tailings. Which novel compounds will be created? How will they affect existing beings and ecologies? To what degree can we interfere in such entanglements, especially when they become toxic to vulnerable populations? In most cases, we believe, such an exercise would reveal that the risks of such future entanglements, especially in the form of toxic geosymbioses, outpace their current or imagined benefits. Instead of providing something like an "ecological license to operate" for new projects, the recognition of the dangers embedded in symbiopower should lead us toward a gradual de-escalation of our relentless thirst for extraction.

In a situation of imminent ecological collapse, when most prognoses only give us a few years to avert utter disaster, a growing recognition of symbiopower appears as a practical and moral imperative. To start seeing extractive projects as producers of multiple entities, both pure and valuable "resources" *and* geosymbioses. To understand that many geosymbioses are truly *indifferent* to human desires, proliferating in ways utterly disconcerting and frequently damaging to human beings and multiple other nonhumans. To recognize that our capacity to control such geosymbioses would always be *limited*, that they can easily turn into monsters against which—quoting again from Cuevas—"you have no options other than entrusting yourself to the saints, nothing more." To accept that our best way to engage with such geosymbioses is through a *politics of weakness*, becoming careful parasites on a living earth that is—in the end—utterly strange to us.

From a standpoint of geosymbiosis and a politics of weakness, spaces such as Carén start to appear not as largely unseen permanent dumps. Neither are they aberrations, mistakes that must be corrected through blaming and remediation. Carén, instead, emerges as engaged in the "permanent recreation of the world" necessary for humans (and many other living beings) to survive the onset of the Anthropocene.[14] Carén is a world of gray and green, a space in which multiple living beings have precariously managed to entangle in vital ways with anthropogenic compounds. It is in these worlds that some tentative possibilities for a common life in the Anthropocene could emerge.

NOTES

PREFACE

1. As stated in Massai and Miranda, "La mitad de la convención."

INTRODUCTION

1. It is important to make a clarification about the name Carén. Originally, the basin and the water stream crossing it were both called Carén. When CODELCO started building up the dam in the 1980s, as we will see in the next chapter, it simply adopted the same name for it. In this book we use "Carén" to refer to Embalse Carén (the dam), Estero Carén (the creek), and Carén (the whole basin). Most of the people participating in this study called the mine either "El Teniente" or simply "Teniente." We have opted for the second option, mostly for the sake of simplicity.

2. Cannell et al., "Geology, Mineralization, Alteration, and Structural Evolution of the El Teniente Porphyry Cu-Mo Deposit," 979.

3. Kossoff et al., "Mine Tailings Dams," 230.

4. From Merriam-Webster dictionaries, www.merriam-webster.com/dictionary/tailings.

5. For a discussion of the different historical meanings of the terms in English-speaking areas see Lawrence and Davies, "Sludge Question," 4.

6. Lottermoser, *Mine Wastes*, 154.

7. Smuda et al., "Geochemistry and Stable Isotope Composition of Fresh Alkaline Porphyry Copper Tailings," 66.

8. Hansen, Yianatos, and Ottosen, "Speciation and Leachability of Copper in Mine Tailings from Porphyry Copper Mining" (small particle size); and Dold, "Evolution of Acid Mine Drainage Formation in Sulphidic Mine Tailings" (acid mine drainage).

9. Förstner, Introduction, 1.

10. International Commission on Large Dams (ICOLD), *Tailings Dams Risk of Dangerous Occurrences*, 15.

11. Besides tailings, industrial copper mines produce discards such as waste rock, mine wastewater, smelting slags, and several chemical compounds (in solid, liquid, and gas forms) used in mineral processing. Hayes et al., "Surficial Weathering of Iron Sulfide Mine Tailings under Semi-Arid Climate," 240 ("comprise[s] the world's largest industrial waste stream"); and Kossoff et al., "Mine Tailings Dams," 230 ("fundamental Earth-shaping geological processes").

12. Smuda et al., "Geochemistry and Stable Isotope Composition of Fresh Alkaline Porphyry Copper Tailings," 63.

13. International Commission on Large Dams (ICOLD), *Tailings Dams Risk of Dangerous Occurrences*, 10.

14. Valenta et al., "Re-Thinking Complex Orebodies."

15. If maintaining the stability of such massive infrastructures when the mine is functional were not challenging enough, dams need to maintain their integrity for decades, even centuries, after the mine has closed. Due to lax regulations and general lack of interest, closure regularly means the utter abandonment of the site, a situation that rapidly causes serious socio-environmental problems such as acid mine drainage. To this we must add the growing risk of utter collapse of the dam due to lack of maintenance and surveillance.

16. Kemp, Owen, and Lèbre, "Tailings Facility Failures in the Global Mining Industry."

17. For overviews see Ballard and Banks, "Resource Wars"; Jacka, "Anthropology of Mining"; Godoy, "Mining"; and Bridge, "Contested Terrain."

18. Jenkins, "Corporate Social Responsibility and the Mining Industry," 24.

19. Gudynas, "Extractivismos: El concepto, sus expresiones y sus múltiples violencias," 62.

20. Svampa, *Neo-Extractivism in Latin America*, 6.

21. Acosta, "Extractivism and Neoextractism: Two Sides of the Same Curse," 69.

22. Boisier et al., "Anthropogenic and Natural Contributions to the Southeast Pacific Precipitation Decline and Recent Megadrought in Central Chile."

23. Klubock, *Contested Communities*.

24. Carrasco, *Embracing the Anaconda*.

25. Golub, *Leviathans at the Gold Mine*; Welker, *Enacting the Corporation*; and Rajak, *In Good Company*.

26. Riofrancos, "Extractivismo Unearthed," 20.

27. Haraway, *Staying with the Trouble*.

28. Dold et al., "Low Molecular Weight Carboxylic Acids in Oxidizing Porphyry Copper Tailings," 2515.

29. Labban, "Deterritorializing Extraction," 567.

30. Tarr, *Search for the Ultimate Sink*.

31. For an exhaustive history of the movement see Sapp, *Evolution by Association*. Haraway, *Companion Species Manifesto*; Haraway, *Staying with the Trouble*; Hird,

Origins of Sociable Life; Kirksey, *Emergent Ecologies*; Helmreich, *Alien Ocean*; and Lorimer, *Probiotic Planet*.

32. Margulis and Sagan, *Slanted Truths*, xxi.

33. Margulis, *The Symbiotic Planet*, 43.

34. Gilbert, "Towards a Developmental Biology of Holobionts," 15.

35. Haraway, *Companion Species Manifesto*.

36. Gilbert et al., "Symbiosis as a Source of Selectable Epigenetic Variation," 672.

37. Moczek, "Re-Evaluating the Environment in Developmental Evolution," 2.

38. Perry et al., "Defining Biominerals and Organominerals" (parts of its structure); and Gadd, "Metals, Minerals and Microbes," 612 ("cycling of metals").

39. Caldwell and Caldwell, "Calculative Nature of Microbe–Mineral Interactions," 261. It is relevant to note that this article, which is so central to our argument, has received only eight citations in the more than fifteen years that has passed since its publication, according to Google Scholar. To claim that it has been even mildly influential is a gross overstatement.

40. Haferburg and Kothe, "Microbes and Metals," 462 ("affect and are affected in return"); and Caldwell and Caldwell, "Calculative Nature of Microbe–Mineral Interactions," 260 ("biological and geological innovation").

41. Schuller, "Microbial Self," 53.

42. Sapp, *Evolution by Association*.

43. As Lorimer comments, with not a little degree of irony, commonly symbiosis theory "provides an appealing worldview for liberal left academics concerned with social inequality and ecological degradation" (*Probiotic Planet*, 14).

44. Sapp, *Evolution by Association*, xv.

45. Daston, *Against Nature*, 4.

46. Leung and Poulin, "Parasitism, Commensalism, and Mutualism," 107.

47. Peacock, "Symbiosis in Ecology and Evolution," 266.

48. In biology, this situation is sometimes referred as "pathogenic parasitism" or a situation in which "emergent or mutant parasite overwhelms the defences of its host, destroying both the host and sometimes itself in the process" (Peacock, "Symbiosis in Ecology and Evolution," 226).

49. Cockerham and Shane, *Basic Environmental Toxicology*, 23.

50. Rozman and Doull, "Dose and Time as Variables of Toxicity," 170.

51. Balayannis and Garnett, "Chemical Kinship."

52. Murphy, "Alterlife and Decolonial Chemical Relations"; Yusoff, *Billion Black Anthropocenes or None*; Povinelli, *Between Gaia and Ground*; and Roberts, "What Gets Inside."

53. As we could imagine, for example, in the case of an addict to drugs such as heroine (relief from other needs). Roberts, "What Gets Inside," 613.

54. Steidinger and Bever, "Host Discrimination in Modular Mutualisms."

55. Povinelli, *Between Gaia and Ground*, 128.

56. Van Dooren, Kirksey, and Munster, "Multispecies Studies," 2.

57. Ogden, Hall, and Tanita, "Animals, Plants, People, and Things," 7.

58. Kirksey and Helmreich, "Emergence of Multispecies Ethnography," 545.

59. Clark and Yusoff, "Geosocial Formations and the Anthropocene," 15.

60. Clark, *Inhuman Nature*, xvi.

61. Clark and Yusoff, "Geosocial Formations and the Anthropocene," 16.

62. Povinelli, *Geontologies*; and Yusoff, *Billion Black Anthropocenes or None*.

63. Povinelli, *Geontologies*, 19.

64. Yusoff, "Geologic Life."

65. Shotwell, *Against Purity*.

66. Tsing, *The Mushroom at the End of the World* (live among the ruins); and Shapiro, "Attuning to the Chemosphere"; Shapiro and Kirksey, "Chemo-Ethnography"; Papadopoulos, "Chemicals, Ecology and Reparative Justice"; and Balayannis and Garnett, "Chemical Kinship."

67. Liboiron, Tironi, and Calvillo, "Toxic Politics," 342 ("burdens of harm as victims"); and Murphy, "Alterlife and Decolonial Chemical Relations," 496.

68. Tuck, "Suspending Damage," 342.

69. For two recent collections see Penfield and Montoya, Introduction; and Vindal Ødegaard and Rivera Andía, *Indigenous Life Projects and Extractivism*.

70. Ulloa, "Feminismos territoriales en América Latina."

71. Viveiros de Castro, *Cannibal Metaphysics*; Kohn, *How Forests Think* (bios); and de la Cadena, *Earth Beings* (geos).

72. Gómez-Barris, *Extractive Zone*, xx.

73. In which waste is mostly "matter out of place" (Douglas, *Purity and Danger*) or "a concept out of order" (Gille, *From the Cult of Waste to the Trash Heap of History*, 23), and hence defined by its pure negativity, something "not to be seen, therefore as merely the material product of industrial development, an expression of monstrous energetic excess" (Cooper, "Recycling Modernity," 1120).

74. Kolbert, *Under a White Sky*, 22.

75. Hadfield and Haraway, "Tree Snail Manifesto," 232.

76. Ballard and Banks, "Resource Wars," 289.

77. Coumans, "Occupying Spaces Created by Conflict" (both critics and facilitators); and Ballard and Banks, "Resource Wars," 289.

78. Welker, *Enacting the Corporation*, 10.

79. Golub, *Leviathans at the Gold Mine*; Welker, *Enacting the Corporation*; and Rajak, *In Good Company*.

80. Tsing, *Mushroom at the End of the World*, 213.

81. Tsing, Mathews, and Bubandt, "Patchy Anthropocene."

82. Ureta, "Selling the Sociotechnical Sublime."

83. Mol, "Ontological Politics."

84. The public office preceding the current Ministry for the Environment.

85. Such exceptionality is not only given to the multiple geosymbioses emerging at Carén, or the public status of CODELCO, as discussed, but also to the characteristics of Teniente as a whole, as a long-running (in contrast with new mining projects, regarding which conflicts are more frequent), underground mine (in contrast with an open-cast one, which tends to be much more environmentally damaging),

located in a secluded area on which there are currently no indigenous ownership claims or even the close proximity of local communities.

86. Tsing, *Mushroom at the End of the World.*

CHAPTER 1. RESIDUALISM

1. "La cuenca del estero Carén," *Semanario El Teniente,* June 6, 1986, 5.

2. Novoa, *La batalla por el cobre,* 121.

3. This was evident in the selection of the central square of Rancagua, the closest city to the mine, as the place for president Allende to give the speech on July 11, 1971, celebrating the nationalization of the copper industry.

4. Allende, "Historical Constraints to Privatization," 76.

5. Despite positioning itself completely at odds with the previous socialist administration, the military shared a "nationalist vision of natural resources as strategic assets that must remain in state control and avoid their appropriation or exploitation by foreign companies," especially a sector as relevant to the nation's economy as copper mining. Nazer, "Nacionalización y privatización del cobre chileno 1971–2002."

6. Allende, "Historical Constraints to Privatization," 71.

7. Allende, "Historical Constraints to Privatization," 77.

8. "El tranque Carén entra en operaciones," *Semanario El Teniente,* November 21, 1986, 1.

9. Nem Singh, "Who Owns the Minerals?," 239.

10. Transition Minerals Tracker, https://trackers.business-humanrights.org/transition-minerals/.

11. Seen at "El cobre y la energía solar," n.d., www.codelco.com/el-cobre-y-la-energia-solar/prontus_codelco/2011-02-17/152212.html (accessed November, 26, 2021).

12. Pignarre and Stengers, *Capitalist Sorcery.*

13. Lagos, *Reflexiones sobre la nacionalización del cobre chileno,* 12.

14. Baros, *El Teniente.*

15. Hiriart, *Braden, historia de una mina.*

16. Millán, *La minería metálica en Chile en el siglo XIX.*

17. Blake-Coleman, *Copper Wire and Electrical Conductors,* xiii.

18. Sheller, *Aluminum Dreams.*

19. Lecain, "Copper and the Evolution of Space in High Modernist America and Japan," 177.

20. Olofsson, "Imagined Futures in Mineral Exploration," 2.

21. Lynch et al., "History of Flotation Technology," 70.

22. Branagan, "Seeking Hidden Millions," 1.

23. Mouat, "Development of the Flotation Process," 13.

24. Lynch et al., "History of Flotation Technology," 69.

25. Sulman, Picard, and Ballot, British Patent 7,803, 3.

26. Mouat, "Development of the Flotation Process," 15.

27. Birrell, "Development of Mining Technology in Australia 1801–1945."

28. Mouat, "Development of the Flotation Process," 25.

29. Frodeman, *Geo-Logic*.

30. It is no accident that two of the companies more closely involved in flotation's development and patenting, Broken Hill Proprietary (currently known as BHP) and the Zinc Corporation (currently known as Rio Tinto), became with the passage of time the two largest mining corporations in the world.

31. Mouat, "Development of the Flotation Process," 4.

32. Mumford, *Technics and Civilization*, 74.

33. *Diario Oficial*, July 13, 1905, 2897. This is the Chilean state's official publication on legislative and commercial issues.

34. For example, in the United States, as stated in the introduction to a 1916 book summarizing recent developments regarding flotation, "In 1912 the flotation process had hardly won a foothold in the United States; today fully 50.000 tons of ore is being treated daily by the frothing or bubble-levitation method" (Rickard, *Flotation Process*, 3). By the end of the decade, "the flotation method of ore treatment was in use in virtually every mining region of the world" (Mouat, "Development of the Flotation Process," 4).

35. Fuenzalida, *El Trabajo i la vida en el mineral "El Teniente"*, 434.

36. Bravo, "Inversiones norteamericanas en Chile," 798.

37. Hiriart, *Braden, historia de una mina*.

38. Bravo, "Inversiones norteamericanas en Chile," 804.

39. Capital that resulted, in practice, in relinquishment of the property of the mine in favor of the Guggenheim brothers, Braden remaining as its CEO.

40. Parsons, *Porphyry Coppers*, 444.

41. After reading the literature on the topic, it is not clear who had the original idea of testing flotation at Teniente. On the one hand, although based in Chile, Chiapponi was well aware of the new developments in the mining industry worldwide, being a subscriber to the latest publications on the matter and traveling frequently to technical fairs (Millán, *La minería metálica en Chile en el siglo XX*, 17), so he easily could have heard about this procedure. Braden, on the other hand, due to its previous works with large mining corporations in the United States, "was well placed to appreciate the main changes that the copper industry was experiencing" (Bravo, "Inversiones norteamericanas en Chile," 798). Probably the answer, as usually happens with technological innovation, lies somewhere in between, with the two men (and probably other people and things) being involved in the process. Several other details of the story, as recounted by the sources collected, are murky, so we have reconstructed it as best as we could.

42. Diaz, "El mineral de 'El Teniente': Interesante informe de injeniero consultor," 44 (encouraging results); and Douglass and Colley, "Metallurgical Operations at the Bradlem Copper Co.," 315 ("amenable to concentration").

43. Parsons, *Porphyry Coppers*, 154.

44. Fuenzalida, *El trabajo i la vida en el mineral "El Teniente"*, 434.

45. Folchi, "Historia ambiental de las labores de beneficio en la minería del cobre en Chile, siglos XIX y XX," 349.

46. Bravo, "Inversiones norteamericanas en Chile," 806.

47. Elliot, *History of Nevada Mines Division*, 50.

48. Arboleda, *Planetary Mine*.

49. Davies, "Tailings Impoundment Failures?"

50. Boudia et al., "Residues," 169.

51. Folchi, "Historia ambiental de las labores de beneficio en la minería del cobre en Chile, siglos XIX y XX," 268.

52. In the case of Teniente such reprocessing has been a reality since 1989, when a company called Minera Valle Central (http://mineravallecentral.cl/) started its operations, focusing on reprocessing tailings from former dams in order to extract copper, enacting in practice a mine within a mine.

53. Something critical given the proximity of Altos de Cantillana, a biodiversity hotspot with more than 40 percent of endemic species.

54. Actually, as one engineer involved in the process told us, he was coerced to sell by the military authorities in power, under the threat that his land would be confiscated if he didn't agree to a selling price with CODELCO after a certain date.

55. Gastó et al., *Plan de ordenación territorial Hacienda Ecológica Los Cobres de Loncha*.

56. The collage was made by the authors with pictures taken from the Flickr page "Añoranzas de Loncha" (Yearnings for Loncha), made by former inhabitants of the hacienda. More images are available at www.flickr.com/photos/145041602@N04/albums/72157675413811281/.

CHAPTER 2. CARP, ALGAE, DRAGON

1. Ureta, "Caring for Waste."

2. Speight, *Environmental Organic Chemistry for Engineers*, 274.

3. Lottermoser, *Mine Wastes*, 157.

4. Deleuze and Guattari, *Thousand Plateaus*.

5. Braun, "Producing Vertical Territory"; Frodeman, *Geo-Logic*; and Yussoff, *Billion Black Anthropocenes or None*.

6. Povinelli, *Geontologies*.

7. Hird, "Waste, Landfills, and an Environmental Ethic of Vulnerability," 106.

8. Smithson, "Sedimentation of the Mind."

9. Smithson, "Sedimentation of the Mind," 82.

10. Johnson, "Fly-Fishing for Carp as a Deeper Aesthetics," 195.

11. Nagy and Johnson, Introduction, 17.

12. Haraway, *Companion Species Manifesto*, 2003.

13. Parra, "Estado de conocimiento de las algas dulceacuícolas de Chile (excepto Bacillariophyceae)."

14. Paerl et al., "Harmful Freshwater Algal Blooms, with an Emphasis on Cyanobacteria," 77.

15. Douglas, *Purity and Danger*.

16. Nagy and Johnson, Introduction, 6.

17. Povinelli, *Geontologies*.

18. Lippincott, "The Unnatural History of Dragons," 20.

19. Taussig, *Devil and Commodity Fetishism in South America*; Nash, *We Eat the Mines and the Mines Eat Us*; and Gudynas, "El petróleo es el excremento del diablo. Demonios, satanes y herejes en los extractivismos," 146.

20. Taussig, *Devil and Commodity Fetishism in South America*, 147.

21. Clark, *Inhuman Nature*, xi.

22. Cohen, *Monster Theory*, 7.

23. Povinelli, *Geontologies*.

24. Such fears seemed to be confirmed on the night of February 27, 2010, when a massive earthquake struck central Chile. As Carvallo told us, he and his family ran to the nearest hill, fleeing what they feared was the imminent collapse of the dam (but that, luckily, did not happen).

25. A vitality that is greatly helped by the earth itself, as central Chile is one of the most geoseismically active regions on earth, experiencing some of the most devastating earthquakes ever recorded, such as the one in February 2010 with a magnitude of 8.8mw and whose epicenter was only 200 kilometers south of Carén.

26. Frodeman, *Geo-Logic*, 36.

27. Gabrys, "Sink," 670.

28. Hird, "Waste, Landfills, and an Environmental Ethic of Vulnerability," 120.

CHAPTER 3. HAPPY COEXISTENCE

1. "El Embalse Caren y el beneficio de sus aguas claras," n.d., www.codelco.com /el-embalse-caren-y-el-beneficio-de-sus-aguas-claras/prontus_codelco/2011-02-18 /094641.html (accessed December 4, 2018).

2. Henke, "Making a Place for Science," 484.

3. Alagona, "Sanctuary for Science," 5.

4. Gieryn, "Three Truth Spots."; and Diser, "Laboratory versus Farm," 46.

5. Alagona, "A Sanctuary for Science"; Diser, "Laboratory versus Farm"; Gieryn, "Three Truth Spots"; Kirsch, "Ecologists and the Experimental Landscape"; Kohler, *Landscapes and Labscapes*; and Minella, "A Pattern for Improvement."

6. Maat, "History and Future of Agricultural Experiments," 187.

7. Kohler, "Labscapes," 495.

8. Consultores en Ingeniería Civil y Arquitectura (CICA), *Uso agropecuario de aguas, efluentes de relaves Carén*.

9. Rheinberger, *Toward a History of Epistemic Things* ("experimental system"); and Stone, "Agriculture as Spectacle."

10. Vila et al., *Análisis crítico de los estudios de evaluación de impacto ambiental de los efluentes del Embalse Carén*, 12.

11. Puig de la Bellacasa, "Making Time for Soil," 698.

12. Craig, "Perceiving Change and Knowing Nature," 87.

13. Ureta, "Baselining Pollution"; and Ureta, Lekan, and Graf von Hardenberg, "Baselining Nature."

14. According to Rheinberger, most experimental systems are composed of a variable mixture of epistemic things and technical objects. Epistemic things, on the one hand, are "that badly defined something that is the very target of a particular experimental research endeavor ... [embodying] what one does not yet exactly know" (*Toward a History of Epistemic Things*, 21). Technical objects, on the other hand, are "the instruments, apparatus, and other devices enabling and at the same time bounding and confining the assessment of the epistemic things under investigation" (21). In contrast with epistemic things, technical objects are fairly rigid objects; "their rigidity and specificity is necessary to keep the vagueness of the epistemic objects limited and to confine their criticality" (21).

15. Referring to "models for specific phenomena, to be investigated through a particular discipline or perspective with its accompanying set of techniques and practices." Ankeny and Leonelli, "What's So Special about Model Organisms?," 319.

16. Consultores en Ingeniería Civil y Arquitectura (CICA), *Uso agropecuario de aguas, efluentes de relaves Carén*, 113.

17. Consultores en Ingeniería Civil y Arquitectura (CICA), *Uso agropecuario de aguas, efluentes de relaves Carén*, 2.

18. Consultores en Ingeniería Civil y Arquitectura (CICA), *Uso agropecuario de aguas, efluentes de relaves Carén*, 2.

19. Delpiano, "Reuso agrícola de aguas claras de relaves mineros," 93.

20. Consultores en Ingeniería Civil y Arquitectura (CICA),*Uso agropecuario de aguas, efluentes de relaves Carén*, 8.

21. Vila et al., *Análisis crítico de los estudios de evaluación de impacto ambiental de los efluentes del Embalse Carén*.

22. Ankeny and Leonelli, "What's So Special about Model Organisms?"

23. Vila et al., *Análisis crítico de los estudios de evaluación de impacto ambiental de los efluentes del Embalse Carén*, 15.

24. Rheinberger, *Toward a History of Epistemic Things*, 21.

25. Vila et al., *Análisis crítico de los estudios de evaluación de impacto ambiental de los efluentes del Embalse Carén*, 14.

26. This was heightened by the consultants' incapacity to produce scientific publications from the data produced at Loncha. Although several draft papers were written and (supposedly) sent to peer-reviewed journals, none was ever accepted for publication.

27. Those negative effects were usually connected to the high amounts of molybdenum in their water and food, causing a disease known as hypocuprosis, a substantial decrease in copper levels in the bloodstream of animals, with several negative

health effects. The concentrations of minerals in water were higher in summer, given increased evaporation.

28. For example, several of the participants in this new round of studies told us that they regularly found extremely high levels of magnesium, a well-known toxicant, in most of their water samples. However, as magnesium was not considered in current Chilean regulation on irrigation water quality, those results were largely dismissed.

29. Combes, *Parasitism*.

30. Frickel et al., "Undone Science."

CHAPTER 4. PARASITISM

1. Ministerio de Obras Públicas (MOP), "DS 609/98 Norma de emisión para la regulación de contaminantes asociados a las descargas de residuos industriales líquidos a sistemas de alcantarillado."

2. Boudia et al., "Residues," 170.

3. Since 2010 this task has been carried out by the Superintendencia del Medioambiente, the body in charge of ensuring the observance of Chile's environmental laws, including the norm regulating RIL discharge from the Carén dam. However, a few kilometers downstream from the dam, the water in Carén Creek is controlled again, this time by the DGA, which treats the water flowing through the Carén as a natural source of water, not questioning its origins.

4. For an overview see Bauer, *Siren Song*; and Budds, "Water, Power, and the Production of Neoliberalism in Chile, 1973–2005."

5. Haferburg and Kothe, "Microbes and Metals," 454.

6. Nixon, *Slow Violence and the Environmentalism of the Poor*.

7. Aldunce et al., "Local Perception of Drought Impacts in a Changing Climate."

8. Beltran, *El DS 80/06*, 24.

9. Papadopoulos, *Experimental Practice*, 198.

10. Bauer, *Siren Song*, 92–96.

11. Smith, "Governing Water," 447.

12. Shiva, *Water Wars*, 35.

13. Something quite surprising given the reluctance of CODELCO to formally disseminate such knowledge among locals, as seen.

14. Ministerio de Vivienda y Urbanismo (MINVU), *Informa inicio de un nuevo procedimiento de evaluación ambiental estratégica del Plan Regulador Intercomunal del Lago Rapel*, 3.

15. Ministerio de Vivienda y Urbanismo (MINVU), *Informa inicio de un nuevo procedimiento de evaluación ambiental estratégica del Plan Regulador Intercomunal del Lago Rapel*, 3.

16. Vila et al., "Rapel."

17. For example, the EPA *Framework for Metals Risk Assessment*, 4–8, 4–12 guidelines about heavy metal pollution assessment, the most used policy instrument on

the matter, recognizes five exposure pathways (inhalation, dietary, exposure, drinking, and dermal). The Chilean guidelines, Ministerio de Medio Ambiente (MMA), Corporación de Fomento (CORFO), and Fundación Chile, *Guía metodológica para la gestión de suelos con potencial presencia de contaminantes*, recognize three (inhalation, dermal, ingestion).

18. Papadopoulos, *Experimental Practice*, 199.

19. Papadopoulos, *Experimental Practice*, 253.

20. Ingeniería y Desarrollo de Proyectos (CADE-IDEPE), *Diagnóstico y clasificación de cursos y cuerpos de agua según objetivos de calidad*, 159.

CHAPTER 5. LIFE AGAINST LIFE

1. "El Teniente responde a críticas por incidente ambiental en tranque Carén," press release, April 17, 2006, www.mch.cl/2006/05/16/el-teniente-responde-a-criticas-por-incidente-ambiental-en-tranque-caren/.

2. Perrow, *Normal Accidents*.

3. The so-called Decreto Supremo 80 (Supreme Decree 80) allows discharging water from the Carén dam with higher levels of molybdenum and sulfate than the regular law, the Supreme Decree 90. So, while Teniente can reach maximum concentrations of up to 1.6 ppm (parts per million) in molybdenum and 2,000 ppm in sulfate in the water discharged into Carén Creek, the rest of the Chilean enterprises cannot surpass 1 ppm in molybdenum and 1000 ppm in sulfate.

4. Cámara de Diputados, *Comisión de recursos naturales, bienes nacionales y medio ambiente... Celebrada en lunes 7 de agosto de 2006, de 15:40 a 19:00 horas*, 8.

5. Cámara de Diputados, *Comisión de recursos naturales, bienes nacionales y medio ambiente... Celebrada en miércoles 23 de agosto de 2006, de 15:40 a 18:25 horas*, 47.

6. Cámara de Diputados, *Comisión de recursos naturales, bienes nacionales y medio ambiente... Celebrada en miércoles 4 de octubre de 2006, de 15:45 a 18:30 horas*, 60.

7. Cámara de Diputados, *Comisión de recursos naturales, bienes nacionales y medio ambiente... Celebrada en miércoles 23 de agosto de 2006, de 15:40 a 18:25 horas*, 47.

8. Servicio Agrícola y Ganadero (SAG), *Incidente ambiental derrame de relaves de cobre embalse Carén*, 8.

9. Cámara de Diputados, *Comisión de recursos naturales, bienes nacionales y medio ambiente... Celebrada en miércoles 4 de octubre de 2006, de 15:45 a 18:30 horas*, 17.

10. The Spanish term *remediar* is frequently used to refer to both ecological restoration and remediation, not distinguishing between them. Hourdequin and Havlick, *Restoring Layered Landscapes*, 1.

11. Hourdequin and Havlick, *Restoring Layered Landscapes*, 1.

12. Hourdequin and Havlick, *Restoring Layered Landscapes*, 1.

13. Eden and Bear, "Models of Equilibrium, Natural Agency and Environmental Change."

14. Fundación Chile, *Remediación de sitio contaminado con relaves del proceso de flotación de cobre*, 2.

15. Cámara de Diputados, *Informe sobre las responsabilidades de la división El Teniente de Codelco Chile*, 11.

16. "Special interest science" or scientific work that is "invested in the cause; their work . . . [being] partial rather than neutral or objective" (Barandarian, *Science and Environment in Chile*, 95). Cámara de Diputados, *Comisión de recursos naturales, bienes nacionales y medio ambiente . . . Celebrada en miércoles 4 de octubre de 2006, de 15:45 a 18:30 horas*, 21.

17. Fundación Chile, *Rehabilitación del estero Carén*, 3.

18. Leopold et al., "Procedure for Evaluating Environmental Impact."

19. Intensity (the level of effect of territorial components), extension (the impact's influence area), moment (the time the impact takes to become manifest), persistence (the duration of the impact), and reversibility (the recuperation capacity of the affected environment).

20. Fundación Chile, *Rehabilitación del estero Carén*, 3.

21. Fundación Chile, *Remediación de sitio contaminado con relaves del proceso de flotación de Cobre*, 10.

22. Corporación Nacional del Cobre (CODELCO), *Memoria anual*, 62.

23. This plan was materialized in several measures, both in general and personal terms. Probably the most relevant measure at the individual level was the payment of monetary compensation to most of the affected neighbors. The strategy proved to be highly successful, with most neighbors receiving a rather small amount of money (up to US$5,000) after signing a legal document stating that the payment was "the single and sufficient compensation for the damages and detriments suffered," closing the door on any further legal action against CODELCO.

24. Their campaign's most visible—and shocking to CODELCO personnel—action was to leave by the side of the public road leading to the entrance of Carén the corpse of each dead cow, with a number painted in bright colors on its stomach representing the total number of dead cattle so far. The corpses were accompanied by large handmade billboards stating things such as "CODELCO does not deliver, lies! [It] contaminated the Creek, wells, soils," and "CODELCO good neighbor, thanks for the tailings water, 146 dead cattle in 30 months."

25. A toxic tort is technically described as a procedure intending to "provide post-injury compensation sufficient to restore the injured person to the condition he or she would have been in had the injury not occurred in the first place." Cranor, *Toxic Torts*, 4.

26. Cranor, *Toxic Torts*, 11.

27. Jasanoff, *Science at the Bar*, 122.

28. Cranor, *Toxic Torts*, 14.

29. Boudia and Jas, "Introduction: Science and Politics in a Toxic World"; and Boudia and Jas, *Powerless Science*.

30. Pemberton, "Environmental Victims and Criminal Justice," 72.

31. Matus, Ramírez, and Castillo, "Acerca de la necesidad de una reforma urgente de los delitos de contaminación en Chile," 789.

32. Barandarian, *Science and Environment in Chile.*

33. On February 27, 2010, an abandoned tailings pond located near the town of Las Palmas (300 km south of Santiago) collapsed after an earthquake, falling on a group of houses located close to it. A young family who lived in one of these houses died, crushed by the tailings, a tragedy that received ample coverage in the national media.

34. For an overview see Dudka and Adriano, "Environmental Impacts of Metal Ore Mining and Processing."

35. Zúñiga/Corporación Nacional del Cobre de Chile (2007), Demanda civil por Daño Ambiental, 2007, 4 (ROL C 67381-2007, Primer Juzgado Civil de Melipilla), www.pjud.cl.

36. Actually, beyond the affirmation made in the initial demand, no proof was ever given that the surviving cattle could not be sold in the market.

37. Unsurprisingly, neither the plaintiffs nor the defense aimed at putting the guilt on the dragon; animism does not go over well in court.

38. Deckha, "Initiating a Non-Anthropocentric Jurisprudence"; Braverman, "Law's Underdog"; Bravo, "Una relectura del estatuto jurídico de los animales en el derecho chileno"; and Leiva Ilabaca, "El delito de maltrato animal en Chile."

39. Mussawir, "The Jurisprudential Meaning of the Animal," 93.

40. Zúñiga/Corporación Nacional del Cobre de Chile, Informe de necropsia, 2007, 2 (ROL C 67381-2007, Primer Juzgado Civil de Melipilla), www.pjud.cl.

41. Zúñiga/Corporación Nacional del Cobre de Chile, Reitera documentos como medios de prueba, 2007 (ROL C 67381-2007, Primer Juzgado Civil de Melipilla), www.pjud.cl.

42. Zúñiga/Corporación Nacional del Cobre de Chile (2007), Téngase presente, 2007, 16 (ROL C 67381-2007, Primer Juzgado Civil de Melipilla), www .pjud.cl.

43. Zúñiga/Corporación Nacional del Cobre de Chile, E 973-2018, 2007, 4 (ROL C 67381-2007, Primer Juzgado Civil de Melipilla), www.pjud.cl.

44. Zúñiga/Corporación Nacional del Cobre de Chile, Sentencia Primer Juzgado de Letras de Melipilla, 2007, 15 (emphasis in original) (ROL C 67381-2007, Primer Juzgado Civil de Melipilla), www.pjud.cl.

45. Rabin, "Sociolegal History of the Tobacco Tort Litigation."

46. Goodie, "Ecological Narrative of Risk and the Emergence of Toxic Tort Litigation," 76.

47. Zúñiga/Corporación Nacional del Cobre de Chile (2007), Sentencia Primer Juzgado de Letras de Melipilla, 2007, 51–52, 56 (ROL C 67381-2007, Primer Juzgado Civil de Melipilla), www.pjud.cl.

48. Paut, *Informe económico-financiero para estimar el monto de la pérdida.*

49. Zúñiga/Corporación Nacional del Cobre de Chile, Informe Económico-Financiero para estimar el monto de la pérdida por Lucro Cesante causado en la masa ganadera y desvalorización del Predio Sector "B" de la Hacienda Pincha, 2007, 12 (ROL C 67381-2007, Primer Juzgado Civil de Melipilla), www.pjud.cl.

1. Corporación Nacional del Cobre (CODELCO), *Plan de cierre faenas mineras*, 154–55.

2. Something similar to Dennis Oppenheim's "Landslide" (1968) or Robert Smithson's "Asphalt Rundown" (1969).

3. Pignarre and Stengers, *Capitalist Sorcery*.

4. Starting from forcing CODELCO into the commitment of continuing to provide water to the basin after the dam has closed.

5. On these topics see Povinelli, *Between Gaia and Ground*.

6. Dalby, "Framing the Anthropocene."

7. Castro, "Who Is Afraid of the Ontological Wolf?," 16.

8. Haraway, *Companion Species Manifesto*, 7.

9. Foucault, *Birth of Biopolitics*; and Povinelli, *Geontologies* (geontopower).

10. As done, for example, in the usage of the term *symbiopolitics* by Helmreich (*Alien Ocean*) or Lorimer (*Probiotic Planet*).

11. Cohen, *Monster Theory*, 4.

12. Hird, *Origins of Sociable Life*; and Wyck, *Primitives in the Wilderness*.

13. A natural contract is a political situation "in which our relationship to things would set aside mastery and possession in favor of admiring attention, reciprocity, contemplation, and respect; where knowledge would no longer imply property, nor action mastery, nor would property and mastery imply their excremental results and origins" (Serres, *Natural Contract*, 38). Latour, *Politics of Nature* (parliament of things); and Povinelli, "Rhetorics of Recognition in Geontopower" (lack any kind of "language").

14. Guattari, *Three Ecologies*, 111.

BIBLIOGRAPHY

Acosta, A. "Extractivism and Neoextractism: Two Sides of the Same Curse." In *Beyond Development Alternative Visions from Latin America*, edited by M. Lang and D. Mokrani, 54–65. Quito: Fundacion Rosa Luxemburgo, 2013.

Alagona, Peter S. "A Sanctuary for Science: The Hastings Natural History Reservation and the Origins of the University of California's Natural Reserve System." *Journal of the History of Biology* 45, no. 4 (2012): 651–80. https://doi.org/10.1007/s10739-011-9298-0.

Aldunce, Paulina, Dámare Araya, Rodolfo Sapiain, Issa Ramos, Gloria Lillo, Anahí Urquiza, and René Garreaud. "Local Perception of Drought Impacts in a Changing Climate: The Mega-Drought in Central Chile." *Sustainability* 9, no. 11 (2017): 2053. https://doi.org/10.3390/su9112053.

Allende, Juan Agustin. "Historical Constraints to Privatization: The Case of the Nationalized Chilean Copper Industry." *Studies in Comparative International Development* 23, no. 1 (1988): 55–84. https://doi.org/10.1007/BF02686999.

Ankeny, Rachel A., and Sabina Leonelli. "What's So Special about Model Organisms?" *Studies in History and Philosophy of Science Part A* 42, no. 2 (2011): 313–23. https://doi.org/10.1016/j.shpsa.2010.11.039.

Arboleda, M. *Planetary Mine: Territories of Extraction under Late Capitalism*. London: Verso, 2020.

Balayannis, A., and E. Garnett. "Chemical Kinship: Interdisciplinary Experiments with Pollution." *Catalyst: Feminism, Theory, Technoscience* 6, no. 1 (2020): 1–10. https://doi.org/10.28968/cftt.v6i1.33524.

Ballard, C., and G. Banks. "Resource Wars: The Anthropology of Mining." *Annual Review of Anthropology* 32 (2003): 287–313.

Barandarian, J. *Science and Environment in Chile: The Politics of Expert Advice in a Neoliberal Democracy*. Cambridge, MA: MIT Press, 2018.

Baros, M. C. *El Teniente: Los hombres del mineral: 1905–1945*. Santiago: Codelco, 1995.

Bauer, Carl J. *Siren Song: Chilean Water Law as a Model for International Reform*. Washington, DC: Resources for the Future, 2004.

Beltran, A. D. *El DS 80/06: Discusión, tramitación y consecuencias de una norma de emisión especial.* Santiago: Centro de Estudios Fiscalia del Medio Ambiente (FIMA), 2007.

Birrell, R. "The Development of Mining Technology in Australia 1801–1945." PhD diss., University of Melbourne, 2005.

Blake-Coleman, B. C. *Copper Wire and Electrical Conductors: The Shaping of a Technology.* Philadelphia, PA: Harwood Academic Publishers, 1992.

Boisier, Juan P., Roberto Rondanelli, René D. Garreaud, and Francisca Muñoz. "Anthropogenic and Natural Contributions to the Southeast Pacific Precipitation Decline and Recent Megadrought in Central Chile." *Geophysical Research Letters* 43, no. 1 (2016): 413–21. https://doi.org/10.1002/2015GL067265.

Boudia, S., A. Creager, S. Frickel, E. Henry, N. Jas, and J. Roberts. "Residues: Rethinking Chemical Environments." *Engaging Science, Technology, and Society* 4 (2018): 165–78.

Boudia, S., and N. Jas. "Introduction: Science and Politics in a Toxic World." In *Toxicants, Health and Regulation since 1945,* 4–25. London: Pickering & Chatto, 2013.

———. *Powerless Science? Science and Politics in a Toxic World.* London: Berghahn Books, 2014.

Branagan, D. "Seeking Hidden Millions—Metallurgists and the Broken Hill Lode." *Journal of Australasian Mining History* 3 (2005): 1–16.

Braun, Bruce. "Producing Vertical Territory: Geology and Governmentality in Late Victorian Canada." *Ecumene* 7, no. 1 (2000): 7–46. https://doi.org/10.1177/096746080000700102.

Braverman, I. "Law's Underdog: A Call for More-Than-Human Legalities." *Annual Review of Law and Social Science* 14, no. 1 (2018): 127–44. https://doi.org/10.1146/annurev-lawsocsci-101317-030820.

Bravo, D. "Una relectura del estatuto jurídico de los animales en el derecho chileno a partir de la vigencia de la ley No 20.380." In *Aproximaciones filosóficas y jurídicas al derecho animal,* 105–17. Antofagasta, Chile: Ediciones Universidad Católica del Norte, 2016.

Bravo, Juan Alfonso. "Inversiones norteamericanas en Chile: 1904–1907." *Revista Mexicana de Sociología* 43, no. 2 (1981): 775–818. https://doi.org/10.2307/3539926.

Bridge, G. "Contested Terrain: Mining and the Environment." *Annual Review of Environmental Resources* 29 (2004): 205–59.

Budds, Jessica. "Water, Power, and the Production of Neoliberalism in Chile, 1973–2005." *Environment and Planning D: Society and Space* 31, no. 2 (2013): 301–18. https://doi.org/10.1068/d9511.

Caldwell, D. E., and S. J. Caldwell. "The Calculative Nature of Microbe–Mineral Interactions." *Microbial Ecology* 47, no. 3 (2004): 252–65. https://doi.org/10.1007/s00248-003-1015-x.

Cámara de Diputados. *Comisión de recursos naturales, bienes nacionales y medio ambiente: 52° período legislativo, 354a legislatura sesión 15a,celebrada en lunes 7 de agosto de 2006, de 15:40 a 19:00 horas.* Valparaiso: Biblioteca del Congreso, 2006.

———. *Comisión de recursos naturales, bienes nacionales y medio ambiente: 52° período legislativo, 354a legislatura sesión 18a, Ordinaria, celebrada en miércoles 23 de agosto de 2006, de 15:40 a 18:25 horas.* Valparaiso: Biblioteca del Congreso, 2006.

———. *Comisión de recursos naturales, bienes nacionales y medio ambiente: 52° período legislativo, 354a legislatura sesión 21a, ordinaria, celebrada en miércoles 4 de octubre de 2006, de 15:45 a 18:30 horas.* Valparaiso: Biblioteca del Congreso, 2006.

———. *Informe sobre las responsabilidades de la división El Teniente de Codelco Chile, por derrame de relaves en el estero Carén y respecto a la tramitación del decreto no. 80, de 2005, del Ministerio Secretaría General de La Presidencia, ante la Contraloría General de la República.* Valparaiso: Cámara de Diputados de Chile, 2006.

Cannell, James, David R. Cooke, John L. Walshe, and Holly Stein. "Geology, Mineralization, Alteration, and Structural Evolution of the El Teniente Porphyry Cu-Mo Deposit." *Economic Geology* 100, no. 5 (2005): 979–1003. https://doi.org/10.2113/gsecongeo.100.5.979.

Carrasco, Anita. *Embracing the Anaconda: A Chronicle of Atacameño Life and Mining in the Andes.* Lanham, MD: Rowman & Littlefield, 2020.

Castro, Eduardo Viveiros de. "Who Is Afraid of the Ontological Wolf? Some Comments on an Ongoing Anthropological Debate." *Cambridge Journal of Anthropology* 33, no. 1 (2015): 2–17. https://doi.org/10.3167/ca.2015.330102.

Clark, N. *Inhuman Nature: Sociable Life on a Dynamic Planet.* London: Sage, 2011.

Clark, N., and K. Yusoff. "Geosocial Formations and the Anthropocene." *Theory, Culture & Society* 34, nos. 2–3 (2017): 3–23. https://doi.org/10.1177/0263276416688946.

Cockerham, Lorris G., and Barbara S. Shane. *Basic Environmental Toxicology.* London: Routledge, 2019.

Cohen, Jeffrey Jerome. *Monster Theory: Reading Culture.* Minneapolis: University of Minnesota Press, 1996.

Combes, C. *Parasitism: The Ecology and Evolution of Intimate Interactions.* Chicago: University of Chicago Press, 2001.

Consultores en Ingeniería Civil y Arquitectura (CICA). *Uso agropecuario de aguas, efluentes de relaves Carén: Resumen 1987–1992, división El Teniente, Codelco-Chile.* Santiago: CICA, 1993.

Cooper, T. "Recycling Modernity: Waste and Environmental History." *History Compass* 8/9 (2010): 1114–25.

Corporación Nacional del Cobre (CODELCO). *Memoria anual.* Santiago: Corporación del Cobre, 2006.

———. *Plan de cierre faenas mineras (Ley No. 20.551), división El Teniente.* Santiago: Corporación Nacional del Cobre, 2014.

Coumans, Catherine. "Occupying Spaces Created by Conflict: Anthropologists, Development NGOs, Responsible Investment, and Mining: With CA Comment by Stuart Kirsch." *Current Anthropology* 52, no. S3 (2011): S29–43. https://doi.org/10.1086/656473.

Craig, R. K. "Perceiving Change and Knowing Nature: Shifting Baselines and Nature's Resiliency." In *Environmental Law and Contrasting Ideas of Nature: A*

Constructivist Approach, edited by K Hirokawa, 87–111. Cambridge: Cambridge University Press, 2014.

Cranor, C. *Toxic Torts: Science, Law, and the Possibility of Justice.* Cambridge: Cambridge University Press, 2006.

Dalby, Simon. "Framing the Anthropocene: The Good, the Bad and the Ugly." *Anthropocene Review* 3, no. 1 (2016): 33–51. https://doi.org/10.1177/2053019615618681.

Daston, L. *Against Nature.* Cambridge, MA: MIT Press, 2019.

Davies, M. "Tailings Impoundment Failures: Are Geotechnical Engineers Listening?" *Geotechnical News* (September 2002): 31–36.

de la Cadena, M. *Earth Beings: Ecologies of Practice across Andean Worlds.* Durham, NC: Duke University Press, 2015.

Deckha, Maneesha. "Initiating a Non-Anthropocentric Jurisprudence: The Rule of Law and Animal Vulnerability under a Property Paradigm." *Alberta Law Review* 50, no. 4 (2013): 783–814.

Deleuze, G., and F. Guattari. *A Thousand Plateaus: Capitalism and Schizophrenia.* London: Athlone Press, 1988.

Delpiano, R. "Reuso agrícola de aguas claras de relaves mineros." In *Memorias del Taller Internacional Organizado por FAO*, 23–34. Arica, Chile: FAO, 1998.

Diaz, I. "El mineral de 'El Teniente': Interesante informe de injeniero consultor." *Boletín de La Sociedad Nacional de Minería* 24, no. 3 (1911): 41–44.

Diser, L. "Laboratory versus Farm: The Triumph of Laboratory Science in Belgian Agriculture at the End of the Nineteenth Century." *Agricultural History* 86, no. 1 (2012): 31–54.

Dold, Bernhard. "Evolution of Acid Mine Drainage Formation in Sulphidic Mine Tailings." *Minerals* 4, no. 3 (2014): 621–41. https://doi.org/10.3390/min4030621.

Dold, Bernhard, David W. Blowes, Ralph Dickhout, Jorge E. Spangenberg, and Hans-Rudolf Pfeifer. "Low Molecular Weight Carboxylic Acids in Oxidizing Porphyry Copper Tailings." *Environmental Science & Technology* 39, no. 8 (2005): 2515–21. https://doi.org/10.1021/es040082h.

Douglas, M. *Purity and Danger: An Analysis of Concepts of Pollution and Taboo.* London: Routledge, 1966.

Douglass, R. E., and B. T. Colley. "Metallurgical Operations at the Bradlem Copper Co." *Engineering & Mining Journal* (February 1916): 315–21.

Dudka, Stanislaw, and Domy C. Adriano. "Environmental Impacts of Metal Ore Mining and Processing: A Review." *Journal of Environmental Quality* 26, no. 3 (1997): 590–602. https://doi.org/10.2134/jeq1997.00472425002600030003x.

Eden, Sally, and Christopher Bear. "Models of Equilibrium, Natural Agency and Environmental Change: Lay Ecologies in UK Recreational Angling." *Transactions of the Institute of British Geographers* 36, no. 3 (2011): 393–407. https://doi.org/10.1111/j.1475-5661.2011.00438.x.

Elliot, R. *History of Nevada Mines Division: Kennecott Copper Corporation.* Reno: University of Nevada Press, 1956.

Environmental Protection Agency (EPA). *Framework for Metals Risk Assessment.* Washington, DC: United States Environmental Protection Agency, 2007.

Folchi, M. "Historia ambiental de las labores de beneficio en la minería del cobre en Chile, siglos XIX y XX." PhD thesis, Universidad de Barcelona, 2006.

Förstner, Ulrich. Introduction to *Environmental Impacts of Mining Activities*, edited by Dr José M. Azcue, 1–3. Berlin: Springer, 1999. https://doi.org/10.1007/978-3-642-59891-3_1.

Foucault, M. *The Birth of Biopolitics: Lectures at the College de France 1978–79*. Basingstoke, UK: Palgrave Macmillan, 2008.

Frickel, S., S. Gibbon, J. Howard, J. Kempner, G. Ottinger, and D. Hess. "Undone Science: Charting Social Movement and Civil Society Challenges to Research Agenda Setting." *Science, Technology, & Human Values* 35, no. 4 (2010): 444–73. https://doi.org/10.1177/0162243909345836.

Frodeman, R. *Geo-Logic: Breaking Ground between Philosophy and the Earth Sciences*. Albany: State University of New York Press, 2003.

Fuenzalida, A. *El trabajo i la vida en el mineral "El Teniente"*. Santiago: Sociedad Imprenta y Litografia Barcelona, 1919.

Fundación Chile. *Rehabilitación del estero Carén—Resumen del estado de avance*. Santiago: Fundación Chile, 2006.

———. *Remediación de sitio contaminado con relaves del Proceso de Flotación de Cobre*. Santiago: Fundación Chile, 2006.

Gabrys, J. "Sink: The Dirt of Systems." *Environment & Planning D: Society and Space* 27 (2009): 666–81.

Gadd, G. "Metals, Minerals and Microbes: Geomicrobiology and Bioremediation." *Microbiology*, 156, no. 3 (2010): 609–43. https://doi.org/10.1099/mic.0.037143-0.

Gastó, J., D. Guzmán, A. Retamal, C. Galvez, P. Rodrigo, T. Tomic, V. Leiva, et al. *Plan de ordenación territorial Hacienda Ecológica Los Cobres de Loncha—Informe final*. Santiago: Programa de Ecología y Medio Ambiente, Facultad de Agronomía e Ingeniería Forestal, Pontificia Universidad Católica de Chile, 2002.

Gieryn, T. "Three Truth Spots." *Journal of History of the Behavioral Sciences* 38, no. 2 (2002): 113–32.

Gilbert, S. "Towards a Developmental Biology of Holobionts." In *Perspectives on Evolutionary and Developmental Biology: Essays for Alessandro Minelli*, edited by G. Fusco, 13–22. Padova, Italy: Padova University Press, 2019.

Gilbert, Scott F., Emily McDonald, Nicole Boyle, Nicholas Buttino, Lin Gyi, Mark Mai, Neelakantan Prakash, and James Robinson. "Symbiosis as a Source of Selectable Epigenetic Variation: Taking the Heat for the Big Guy." *Philosophical Transactions of the Royal Society B: Biological Sciences* 365, no. 1540 (2010): 671–78. https://doi.org/10.1098/rstb.2009.0245.

Gille, Z. *From the Cult of Waste to the Trash Heap of History: The Politics of Waste in Socialist and Postsocialist Hungary*. Bloomington: Indiana University Press, 2007.

Godoy, R. "Mining: Anthropological Perspectives." *Annual Review of Anthropology* 14, no. 1 (1985): 199–217. https://doi.org/10.1146/annurev.an.14.100185.001215.

Golub, A. *Leviathans at the Gold Mine: Creating Indigenous and Corporate Actors in Papua New Guinea*. Durham, NC: Duke University Press, 2014.

Gómez-Barris, M. *The Extractive Zone: Social Ecologies and Decolonial Perspectives.* Durham, NC: Duke University Press, 2017.

Goodie, J. "The Ecological Narrative of Risk and the Emergence of Toxic Tort Litigation." In *Law and Ecology: New Environmental Foundations*, edited by A. Philippopoulos-Mihalopoulos, 65–82. London: Routledge, 2011.

Guattari, F. *The Three Ecologies.* London: Athlone Press, 2000.

Gudynas, E. "El petróleo es el excremento del diablo: Demonios, satanes y herejes en los extractivismos." *Tabula Rasa*, no. 24 (2016): 145–67.

———. "Extractivismos: El concepto, sus expresiones y sus múltiples violencias." *Papeles de Relaciones Ecosociales y Cambio Global* 143 (2018): 61–70.

Hadfield, Michael G., and Donna J. Haraway. "The Tree Snail Manifesto." *Current Anthropology* 60, no. S20 (August 1, 2019): S209–35. https://doi.org/10.1086/703377.

Haferburg, Götz, and Erika Kothe. "Microbes and Metals: Interactions in the Environment." *Journal of Basic Microbiology* 47, no. 6 (2007): 453–67. https://doi.org/10.1002/jobm.200700275.

Hansen, Henrik K., Juan B. Yianatos, and Lisbeth M. Ottosen. "Speciation and Leachability of Copper in Mine Tailings from Porphyry Copper Mining: Influence of Particle Size." *Chemosphere* 60, no. 10 (2005): 1497–1503. https://doi.org/10.1016/j.chemosphere.2005.01.086.

Haraway, D. *The Companion Species Manifesto: Dogs, People and Significant Otherness.* Chicago: Prickly Paradigm Press, 2003.

———. *Staying with the Trouble: Making Kin in the Chthulucene.* Durham, NC: Duke University Press, 2016.

Hayes, Sarah M., Robert A. Root, Nicolas Perdrial, Raina M. Maier, and Jon Chorover. "Surficial Weathering of Iron Sulfide Mine Tailings under Semi-Arid Climate." *Geochimica et Cosmochimica Acta* 141 (2014): 240–57. https://doi.org/10.1016/j.gca.2014.05.030.

Helmreich, S. *Alien Ocean: Anthropological Voyages in Microbial Seas:* Berkeley: University of California Press, 2009.

Henke, Christopher R. "Making a Place for Science: The Field Trial." *Social Studies of Science* 30, no. 4 (2000): 483–511. https://doi.org/10.1177/0306312000300004001.

Hird, M. *The Origins of Sociable Life: Evolution after Science Studies.* London: Palgrave Macmillan, 2009.

———. "Waste, Landfills, and an Environmental Ethic of Vulnerability." *Ethics & the Environment* 18, no. 1 (2013): 105–24.

Hiriart, L. *Braden, historia de una mina.* Santiago: Editorial Andes, 1964.

Hourdequin, M., and D. Havlick, eds. *Restoring Layered Landscapes: History, Ecology, and Culture.* Oxford: Oxford University Press, 2016.

Ingeniería y Desarrollo de Proyectos (CADE-IDEPE). *Diagnóstico y clasificación de cursos y cuerpos de agua según objetivos de calidad—Cuenca del río Rapel.* Santiago: Dirección General de Aguas—Ministerio de Obras Públicas, 2004.

International Commission on Large Dams (ICOLD). *Tailings Dams Risk of Dangerous Occurrences: Lessons Learnt from Practical Experiences.* Paris: Commission Internationale des Grands Barrages, 2001.

Jacka, J. "The Anthropology of Mining: The Social and Environmental Impacts of Resource Extraction in the Mineral Age." *Annual Review of Anthropology* 47 (2018): 61–77.

Jasanoff, S. *Science at the Bar: Law, Science, and Technology in America.* Cambridge, MA: Harvard University Press, 1995.

Jenkins, Heledd. "Corporate Social Responsibility and the Mining Industry: Conflicts and Constructs." *Corporate Social Responsibility and Environmental Management* 11, no. 1 (2004): 23–34. https://doi.org/10.1002/csr.50.

Johnson, P. H. "Fly-Fishing for Carp as a Deeper Aesthetics." In *Trash Animals: How We Live with Nature's Filthy, Feral, Invasive, and Unwanted Species*, edited by K. Nagy, 182–200. Minneapolis: University of Minnesota Press, 2013.

Kemp, Deanna, John R. Owen, and Éléonore Lèbre. "Tailings Facility Failures in the Global Mining Industry: Will a 'Transparency Turn' Drive Change?," *Business Strategy and the Environment* 30, no. 1 (2021): 122–34. https://doi.org/10.1002/bse.2613.

Kirksey, E. *Emergent Ecologies.* Durham, NC: Duke University Press, 2015.

Kirksey, S. Eben, and Stefan Helmreich. "The Emergence of Multispecies Ethnography." *Cultural Anthropology* 25, no. 4 (2010): 545–76. https://doi.org/10.1111/j.1548-1360.2010.01069.x.

Kirsch, S. "Ecologists and the Experimental Landscape: The Nature of Science at the US Department of Energy's Savannah River Site." *Cultural Geographies* 14 (2007): 485–510.

Klubock, T. M. *Contested Communities: Class, Gender, and Politics in Chile's El Teniente Copper Mine, 1904–1951.* Durham, NC: Duke University Press, 1998.

Kohler, R. "Labscapes: Naturalizing the Lab." *History of Science* 40, no. 4 (2002): 473–501. https://doi.org/10.1177/007327530204000405.

———. *Landscapes and Labscapes: Exploring the Lab-Field Border in Biology.* Chicago: University of Chicago Press, 2002.

Kohn, E. *How Forests Think: Toward an Anthropology beyond the Human.* Berkeley: University of California Press, 2013.

Kolbert, E. *Under a White Sky: The Nature of the Future.* New York: Crown, 2021.

Kossoff, D., W. E. Dubbin, M. Alfredsson, S. J. Edwards, M. G. Macklin, and K. A. Hudson-Edwards. "Mine Tailings Dams: Characteristics, Failure, Environmental Impacts, and Remediation." *Applied Geochemistry* 51 (2014): 229–45. https://doi.org/10.1016/j.apgeochem.2014.09.010.

Labban, M. "Deterritorializing Extraction: Bioaccumulation and the Planetary Mine." *Annals of the Association of American Geographers* 104, no. 3 (2014): 560–76. https://doi.org/10.1080/00045608.2014.892360.

Lagos, G. *Reflexiones sobre la nacionalización del cobre chileno.* Santiago: Programa de Investigación en Economía de Minerales, Pontificia Universidad Católica de Chile, 2011.

Latour, B. *Politics of Nature; How to Bring the Sciences into Democracy*. Cambridge, MA.: Harvard University Press, 2004.

Lawrence, Susan, and Peter Davies. "The Sludge Question: The Regulation of Mine Tailings in Nineteenth-Century Victoria." *Environment and History* 20, no. 3 (2014): 385–410. https://doi.org/10.3197/096734014X14031694156448.

Lecain, J. "Copper and the Evolution of Space in High Modernist America and Japan." *Jahrbuch Für Wirtschaftsgeschichte/Economic History Yearbook* 57, no. 1 (2016): 169–86. https://doi.org/10.1515/jbwg-2016-0008.

Leiva Ilabaca, C. "El delito de maltrato animal en Chile: Historia del artículo 291 bis y análisis crítico a la luz del nuevo tipo penal incorporado por la ley 21.020." In *Derecho animal: Teoría y práctica*, edited by M. J. Chible and J. Gallego, 405–26. Santiago: Thompson Reuters, 2018.

Leopold, L., F. Clarke, B. Hanshaw, and J. Basley. "A Procedure for Evaluating Environmental Impact." *Geological Survey Circular* 645 (1971): 1–13.

Leung, T., and R. Poulin. "Parasitism, Commensalism, and Mutualism: Exploring the Many Shades of Symbioses." *Vie et Milieu* 58, no. 2 (2008): 107–15.

Liboiron, M., M. Tironi, and N. Calvillo. "Toxic Politics: Acting in a Permanently Polluted World." *Social Studies of Science* 48, no. 3 (2018): 331–49. https://doi.org/10.1177/0306312718783087.

Lippincott, L. "The Unnatural History of Dragons." *Philadelphia Museum of Art Bulletin* 77, no. 334 (1981): 2–24.

Lorimer, J. *The Probiotic Planet: Using Life to Manage Life*. Minneapolis: University of Minnesota Press, 2020.

Lottermoser, B. *Mine Wastes: Characterization, Treatment, Environmental Impacts*. Berlin: Springer, 2007.

Lynch, A., J. Watt, J. Finch, and G. Harbort. "History of Flotation Technology." In *Froth Flotation. A Century of Innovation*, edited by C. Fuerstenau, G Jameson, and R. Yoon, 65–91. Littleton, CO: SME, 2007.

Maat, H. "The History and Future of Agricultural Experiments." *NJAS-Wageningen Journal of Life Sciences* 57 (2011): 187–95.

Margulis, L. *The Symbiotic Planet: A New Look at Evolution*. London: Phoenix, 1998.

Margulis, L., and D. Sagan. *Slanted Truths: Essays on Gaia, Symbiosis, and Evolution*. New York: Springer-Verlag, 1997.

Massai, N., and B. Miranda. "La mitad de la convención: 77 constituyentes electos provienen de listas que impulsan cambios radicales al sistema." *CIPER Chile*, May 17, 2021, sec. Actualidad. https://www.ciperchile.cl/2021/05/17/la-mitad-de-la-convencion-77-constituyentes-electos-provienen-de-listas-que-impulsan-cambios-radicales-al-sistema/.

Matus, J. P., C. Ramírez, and M. Castillo. "Acerca de la Necesidad de una reforma urgente de los delitos de contaminación en Chile, a la luz de la evolución legislativa del siglo XXI." *Política Criminal* 13, no. 26 (2018): 771–835. https://doi.org/10.4067/S0718-33992018000200771.

Millán, A. *La minería metálica en Chile en el siglo XIX*. Santiago: Editorial Universitaria, 2004.

————. *La minería metálica en Chile en el siglo XX*. Santiago: Editorial Universitaria, 2006.

Minella, T. "A Pattern for Improvement: Pattern Farms and Scientific Authority in Early Nineteenth-Century America." *Agricultural History* 90, no. 4 (2016): 434–58.

Ministerio de Medio Ambiente (MMA), Corporación de Fomento (CORFO), and Fundación Chile. *Guía metodológica para la gestión de suelos con potencial presencia de contaminantes*. Santiago: Ministerio de Medio Ambiente—Gobierno de Chile, 2012.

Ministerio de Obras Públicas (MOP). "DS 609/98 Norma de emisión para la regulación de contaminantes asociados a las descargas de residuos industriales líquidos a sistemas de alcantarillado." *Diario Oficial de La Republica de Chile* (1998).

Ministerio de Vivienda y Urbanismo (MINVU). *Informa inicio de un nuevo procedimiento de evaluación ambiental estratégica del Plan Regulador Intercomunal del Lago Rapel*. Santiago: Ministerio de Vivienda y Urbanismo, Gobierno de Chile, 2018.

Moczek, Armin P. "Re-Evaluating the Environment in Developmental Evolution." *Frontiers in Ecology and Evolution* 3 (2015). https://doi.org/10.3389/fevo.2015.00007.

Mol, A. "Ontological Politics: A Word and Some Questions." In *Actor Network Theory and After*, 74–89. Oxford: Blackwell, 1999.

Mouat, Jeremy. "The Development of the Flotation Process: Technological Change and the Genesis of Modern Mining, 1898–1911." *Australian Economic History Review* 36, no. 1 (1996): 3–31. https://doi.org/10.1111/aehr.361001.

Mumford, L. *Technics and Civilization*. London: Routledge & Kegan, 1934.

Murphy, Michelle. "Alterlife and Decolonial Chemical Relations." *Cultural Anthropology* 32, no. 4 (2017): 494–503. https://doi.org/10.14506/ca32.4.02.

Mussawir, E. "The Jurisprudential Meaning of the Animal: A Critique of the Subject of Rights in the Laws of Scienter and Negligence." In *Law and the Question of the Animal: A Critical Jurisprudence*, edited by Y. Otomo and E. Mussawir, 89–101. London: Routledge, 2013.

Nagy, K., and P. Johnson. Introduction to *Trash Animals: How We Live with Nature's Filthy, Feral, Invasive, and Unwanted Species*, 1–30. Minneapolis: University of Minnesota Press, 2013.

Nash, J. C. *We Eat the Mines and the Mines Eat Us: Dependency and Exploitation in Bolivian Tin Mines*. New York: Columbia University Press, 1993.

Nazer, R. "Nacionalización y privatización del cobre chileno 1971–2002." *Pensamiento Crítico* 4 (2004): 1–15.

Nem Singh, J. "Who Owns the Minerals? Repoliticizing Neoliberal Governance in Brazil and Chile." *Journal of Developing Societies* 28, no. 2 (2012): 229–55.

Nixon, R. *Slow Violence and the Environmentalism of the Poor*. Cambridge, MA: Harvard University Press, 2011.

Novoa, E. *La batalla por el cobre: Comentarios y documentos*. Santiago: Quimantu, 1972.

Ogden, L., B. Hall, and K. Tanita. "Animals, Plants, People, and Things: A Review of Multispecies Ethnography." *Environment and Society: Advances in Research* 4 (2013): 5–24.

Olofsson, Tobias. "Imagined Futures in Mineral Exploration." *Journal of Cultural Economy* 13, no. 3 (2020): 265–77. https://doi.org/10.1080/17530350.2019.1604399.

Paerl, Hans W., Rolland S. Fulton, Pia H. Moisander, and Julianne Dyble. "Harmful Freshwater Algal Blooms, with an Emphasis on Cyanobacteria." *Scientific World* 1 (2001): 76–113. https://doi.org/10.1100/tsw.2001.16.

Papadopoulos, D. "Chemicals, Ecology and Reparative Justice." In *Reactivating Elements. Substance, Actuality and Practice from Chemistry to Cosmology*, edited by D. Papadopoulos, M. Puig de la Bellacasa, and N. Myers, 34–69. Durham, NC: Duke University Press, 2021.

———. *Experimental Practice: Technoscience, Alterontologies, and More-Than-Social Movements.* Durham, NC: Duke University Press, 2018.

Parra, Oscar. "Estado de conocimiento de las algas dulceacuícolas de Chile (excepto Bacillariophyceae)." *Gayana (Concepción)* 70, no. 1 (2006): 8–15. https://doi.org/10.4067/S0717-65382006000100003.

Parsons, A. B. *The Porphyry Coppers.* New York: American Institute of Mining and Metallurgical Engineers, 1933.

Paut, C. *Informe económico-financiero para estimar el monto de la pérdida por lucro cesante causado en la masa ganadera y desvalorización del predio sector "b" de la Hacienda Pincha, ubicado en la localidad de Quilamuta—Loncha, comuna de Alhué. Por un derrame de relaves de CODELCO en abril de 2006.* Technical report presented at trial, Santiago, 2018.

Peacock, Kent A. "Symbiosis in Ecology and Evolution." In *Philosophy of Ecology*, edited by Kevin deLaplante, Bryson Brown, and Kent A. Peacock, 11:219–50. Handbook of the Philosophy of Science. Amsterdam: North-Holland, 2011. https://doi.org/10.1016/B978-0-444-51673-2.50009-1.

Pemberton, A. "Environmental Victims and Criminal Justice: Proceed with Caution." In *Environmental Crime and Its Victims: Perspectives within Green Criminology*, edited by T. Spapens, R. White, and M. Kluin, 63–86. Burlington NJ: Ashgate, 2014.

Penfield, Amy, and Ainhoa Montoya. "Introduction: Resource Engagements; Experiencing Extraction in Latin America." *Bulletin of Latin American Research* 39, no. 3 (2020): 287–89. https://doi.org/10.1111/blar.13069.

Perrow, C. *Normal Accidents: Living with High-Risk Technologies.* Princeton, NJ.: Princeton University Press, 1999.

Perry, Randall S., Nicola Mcloughlin, Bridget Y. Lynne, Mark A. Sephton, Joan D. Oliver, Carole C. Perry, Kathleen Campbell, et al. "Defining Biominerals and Organominerals: Direct and Indirect Indicators of Life." *Sedimentary Geology* 201, no. 1 (2007): 157–79. https://doi.org/10.1016/j.sedgeo.2007.05.014.

Pignarre, P., and I. Stengers. *Capitalist Sorcery: Breaking the Spell.* London: Palgrave Macmillan, 2011.

Povinelli, E. *Between Gaia and Ground: Four Axioms of Existence and the Ancestral Catastrophe of Late Liberalism*. Durham, NC: Duke University Press, 2021.

———. *Geontologies: A Requiem to Late Liberalism*. Durham, NC: Duke University Press, 2016.

———. "The Rhetorics of Recognition in Geontopower." *Philosophy & Rhetoric* 48, no. 4 (2015): 428–42.

Puig de la Bellacasa, M. "Making Time for Soil: Technoscientific Futurity and the Pace of Care." *Social Studies of Science* 45, no. 5 (2015). https://doi.org/10.1177/0306312715599851.

Rabin, Robert L. "A Sociolegal History of the Tobacco Tort Litigation." *Stanford Law Review* 44 (1992): 853.

Rajak, D. *In Good Company: An Anatomy of Corporate Social Responsibility*. Stanford, CA: Stanford University Press, 2011.

Rheinberger, Hans-Jorg. *Toward a History of Epistemic Things: Synthesizing Proteins in the Test Tube*. Stanford, CA: Stanford University Press, 1997.

Rickard, T. A., ed. *The Flotation Process*. London: Mining and Scientific Press, 1916.

Riofrancos, Thea. "Extractivismo Unearthed: A Genealogy of a Radical Discourse." *Cultural Studies* 31, nos. 2–3 (2017): 277–306. https://doi.org/10.1080/09502386.2017.1303429.

Roberts, Elizabeth F. S. "What Gets Inside: Violent Entanglements and Toxic Boundaries in Mexico City." *Cultural Anthropology* 32, no. 4 (2017): 592–619. https://doi.org/10.14506/ca32.4.07.

Rozman, Karl K., and John Doull. "Dose and Time as Variables of Toxicity." *Toxicology* 144, no. 1 (2000): 169–78. https://doi.org/10.1016/S0300-483X(99)00204-8.

Sapp, J. *Evolution by Association: A History of Symbiosis*. Oxford: Oxford University Press, 1994.

Schuller, Kyla. "The Microbial Self: Sensation and Sympoiesis." *Resilience: A Journal of the Environmental Humanities* 5, no. 3 (2018): 51–67.

Serres, M. *The Natural Contract*. Ann Arbor: University of Michigan Press, 1995.

Servicio Agrícola y Ganadero (SAG). *Incidente ambiental derrame de relaves de cobre embalse Carén*. Melipilla, Región Metropolitana: Servicio Agrícola y Ganadero, Gobierno de Chile, 2006.

Shapiro, N. "Attuning to the Chemosphere: Domestic Formaldehyde, Bodily Reasoning, and the Chemical Sublime." *Cultural Anthropology* 30, no. 3 (2015): 368–93.

Shapiro, N., and E. Kirksey. "Chemo-Ethnography: An Introduction." *Cultural Anthropology* 32, no. 4 (2017): 481–93.

Sheller, M. *Aluminum Dreams: The Making of Light Modernity*. Cambridge, MA: MIT Press, 2014.

Shiva, V. *Water Wars: Privatization, Pollution, and Profit*. Berkeley, CA: North Atlantic Books, 2016.

Shotwell, A. *Against Purity: Living Ethically in Compromised Times*. Minneapolis: University of Minnesota Press, 2016.

Smith, Henry E. "Governing Water: The Semicommons of Fluid Property Rights." *Arizona Law Review* 50 (2008): 445.

Smithson, R. "A Sedimentation of the Mind: Earth Projects." *Artforum* (September 1968): 82–91.

Smuda, J., B. Dold, J. Spangenberg, and H. Pfeifer. "Geochemistry and Stable Isotope Composition of Fresh Alkaline Porphyry Copper Tailings: Implications on Sources and Mobility of Elements during Transport and Early Stages of Deposition." *Chemical Geology* 256, nos. 1–2 (2008): 62–76. https://doi.org/10.1016/j.chemgeo.2008.08.001.

Speight, J. *Environmental Organic Chemistry for Engineers,*. Amsterdam: Elsevier, 2017.

Steidinger, Brian S., and James D. Bever. "Host Discrimination in Modular Mutualisms: A Theoretical Framework for Meta-Populations of Mutualists and Exploiters." *Proceedings of the Royal Society B: Biological Sciences* 283, no. 1822 (2016): 20152428. https://doi.org/10.1098/rspb.2015.2428.

Stone, G. "Agriculture as Spectacle." *Journal of Political Ecology* 25 (2018): 656–85.

Sulman, H. L., H. F. K. Picard, and J. Ballot. British Patent 7,803. London, filed 1912 and issued 1912.

Svampa, M. *Neo-Extractivism in Latin America: Socio-Environmental Conflicts, the Territorial Turn, and New Political Narratives*. Cambridge: Cambridge University Press, 2019. https://doi.org/10.1017/9781108752589.

Tarr, J. *The Search for the Ultimate Sink: Urban Pollution in Historical Perspective*. Akron, OH: University of Akron Press, 1996.

Taussig, M. *The Devil and Commodity Fetishism in South America*. Chapel Hill: University of North Carolina Press, 2010.

Tsing, A. *The Mushroom at the End of the World: On the Possibility of Life in Capitalist Ruins*. Princeton, NJ: Princeton University Press, 2015.

Tsing, A., A. Mathews, and N. Bubandt. "Patchy Anthropocene: Landscape Structure, Multispecies History, and the Retooling of Anthropology: An Introduction to Supplement 20." *Current Anthropology* 60, no. S20 (2019): S186–97. https://doi.org/10.1086/703391.

Tuck, Eve. "Suspending Damage: A Letter to Communities." *Harvard Educational Review* 79, no. 3 (2009): 409–28. https://doi.org/10.17763/haer.79.3.n0016675 66it3n15.

Ulloa, Astrid. "Feminismos territoriales en América Latina: Defensas de la vida frente a los extractivismos." *Nómadas*, no. 45 (2018): 123–39.

Ureta, S. "Baselining Pollution: Producing 'Natural Soil' for an Environmental Risk Assessment Exercise in Chile." *Journal of Environmental Policy & Planning* 20, no. 3 (2018): 342–55. https://doi.org/10.1080/1523908X.2017.1410430.

———. "Caring for Waste: Handling Tailings in a Chilean Copper Mine." *Environment & Planning A* 48, no. 8 (2016): 1532–48.

———. "Selling the Sociotechnical Sublime: Critical Reflections on Introducing STS to Managers of a Chilean Mining Corporation." *Tapuya: Latin American Science, Technology and Society* 1, no. 1 (2018): 138–52. https://doi.org/10.1080/25729861.2018.1485246.

Ureta, S., T. Lekan, and W. Graf von Hardenberg. "Baselining Nature: An Introduction." *Environment & Planning E: Nature and Space* 3, no. 1 (2020): 3–19.

Valenta, R. K., D. Kemp, J. R. Owen, G. D. Corder, and É. Lèbre. "Re-Thinking Complex Orebodies: Consequences for the Future World Supply of Copper." *Journal of Cleaner Production* 220 (2019): 816–26. https://doi.org/10.1016/j.jclepro.2019.02.146.

van Dooren, T., E. Kirksey, and U. Munster. "Multispecies Studies: Cultivating Arts of Attentiveness." *Environmental Humanities* 8, no. 1 (2016).

Vila, I., M. Contreras, C. de la Maza, M. Bronffman, J. Pizarro, E. Riveros, and J. Castillo. *Análisis crítico de los estudios de evaluación de impacto ambiental de los efluentes del embalse Carén.* Santiago: Dirección General de Aguas—Ministerio de Obras Públicas, 1993.

Vila, I., M. Contreras, V. Montecino, J. Pizarro, and D. Adams. "Rapel: A 30 Years Temperate Reservoir; Eutrophication or Contamination?" *Archiv für Hydrobiologie* 55 (2000): 31–44.

Vindal Ødegaard, Cecilie, and Juan Javier Rivera Andía. *Indigenous Life Projects and Extractivism: Ethnographies from South America.* Berlin Springer Nature, 2019. https://library.oapen.org/handle/20.500.12657/23130.

Viveiros de Castro, E. *Cannibal Metaphysics.* Minneapolis, MN: Univocal Publishing, 2014.

Welker, M. *Enacting the Corporation: An American Mining Firm in Post-Authoritarian Indonesia.* Berkeley: University of California Press, 2014.

Wyck, P. van. *Primitives in the Wilderness: Deep Ecology and the Missing Human Subject.* Albany: State University of New York Press, 1997.

Yusoff, K. *A Billion Black Anthropocenes or None.* Minneapolis: University of Minnesota Press, 2018.

———. "Geologic Life: Prehistory, Climate, Futures in the Anthropocene." *Environment & Planning D: Society and Space* 31 (2013): 779–95.

INDEX

Corporación Nacional del Cobre
(CODELCO) (*continued*)
spill, 23, 45, 48, 50, 65, 69, 74, 86,
89–98, 100–101, 103–5
corporate social responsibility (CSR), 9, 98,
109; CSR 2.0, 20
Cuevas, Raúl (pseud.), 39–41, 44–45,
47–53, 114

damage, 4–7, 10, 13–14, 17, 21, 23, 43, 93–104,
110–13; ecological damage, 94–95, 98;
environmental damage, 96, 104; damage-
centered narratives, 17, 18; damaged
ecology, 105; toxic damage, 23, 101
Department of Environment Health,
Ministry of Health, 101
Dirección General de Aguas (DGA), 64,
80–81; DGA report, 64
discharge tower, 41, 49–51, 53, 84
dragon, 19, 22, 39, 47–54, 88–89, 91–93, 105,
108, 110, 112

ecology: ecological disaster, 94; ecological
purity, 23, 105; ecological restoration,
94; ecological ruination, 106
Empresa Nacional de Electricidad Sociedad
Anónima (ENDESA), 72, 74, 77, 81, 87
Espejo, Fernando (pseud.), 44, 45, 47, 54
Estación Experimental Loncha/Loncha
Experimental Station, 8, 22, 58–59, 61,
63, 65, 67, 68–70, 72, 92
eternal experimentation, 69, 81
eternal irresolution, 81
experimental stations, 59
extractivismo/extractivism, 6, 107; alternatives to, 17; extractivismo literature of
mining, 10
extractivismo narratives, 9, 109

Figueroa, Luisa (pseud.), 67–68
flotation, 21, 29–30, 33; flotation in Chile,
31; flotation at Teniente, 32–35, 38
Fuentes, Nelson (pseud.), 84–86
Fundación Chile, 95–97, 101–2

Galindo, Héctor (pseud.), 68–69, 71, 81
geontopower/geontologies, 16, 50, 111

geos, 16–17; geos literature, 18; nonhuman
geos, 16
geosymbiosis, 12–13, 20, 22, 47, 67, 81, 107,
109, 112–14; geosymbiont, 113; toxic
geosymbiosis, 22
gravity separation, 28
gray worlds, 3

Hacienda Loncha, 36, 59
hacienda system, 36, 110
happy coexistence, 22, 58, 61, 65, 67, 69–70,
74, 78, 95, 106, 112
Haraway, Donna, 10
harm: emergent harm (daño emergente),
100; loss of profit (lucro cesante), 100
Hermosilla, Fernando (pseud.), 50–53
holobiome, 11
holo(geo)biome, 12
human right to leisure, 87

infernal alternative, 26, 110
Inostroza, Sergio (pseud.), 50–51
inverse resistance, 74, 81, 86, 88

Kennecott Mining corporation, 33

Laguna de los Patos, 95
La Pobrecita, 56–57, 59, 67–68, 70, 80
Leppe, Raúl (pseud.), 91
liquefaction, 5, 33
Loncha, 73, 76
López, Pedro (pseud.), 94–95
low-grade deposits, 29–30

Margulis, Lynn, 11
Márquez, Rodolfo (pseud.), 82–85
Matilda, 56, 58, 66, 68
megadrought, 8, 10, 72, 76, 78, 81–82,
86–87, 108–10
Ministry of Housing and Urbanism, 83
molybdenum, 40, 44–45, 60–61, 63, 67;
concentration of, 47, 71; particles of, 46;
treatment plant (PAMo), 45, 47, 58,
71–72
Monsalve (pseud.): Monsalve family, 98,
103, 106; Ricardo Monsalve (pseud.),
76–79, 93–94, 98

National Institute for Normalization, 80
normal accidents, 90, 92

parasitic agriculture, 22, 68, 70, 77, 81
Pérez, Daniel (pseud.), 92
Pincha, 72–74, 76, 82
Pinochet, Augusto, 25; Pinochet's military
    dictatorship, 72
Pinto, Juan (pseud.), 52–53
politics of weakness, 23, 112–14
porphyry copper deposits, 28–29
Povinelli, Elizabeth, 14, 16
purity: chemical purity, 94; environmental
    purity, 77; standards of purity, 80;
    ultimate purity, 110

Quilamuta, 72–73
Quiñones, Gustavo (pseud.), 60

Rapel: Rapel hydroelectric plant, 72; Rapel
    Lake, 72–74, 76–77, 83–84, 86–87
residualism, 14, 21, 22, 24, 26, 38, 40–41,
    49–50, 58, 70, 109–10
residue: liquid industrial residues (RILs),
    71; mining residues, 26, 109
riparian rights, 77–78

sedimentation: deep-time sedimentation,
    44, 49, 53; desedimentation, 54; double
    sedimentation, 41; naturalization, 40;
    normalization, 40; sedimentations
    lagoons, 40; sedimentation of extrac-
    tion, 47; sedimentation process, 39–40
special interest science, 96

symbiopower, 18, 23, 107, 111–14
symbioses: biological symbioses, 12, 14;
    commensalism, 13; mutualism, 13;
    parasitism, 13; symbiont, 12–14
Suez Degremont, 45
sulfates, 60–61, 63, 67

tailings: tailings agencies, 92; tailings-as-
    dragons, 53; tailings canal, 2, 9, 11, 24,
    39; tailings dams/tailings ponds, 2, 5, 6,
    7, 18, 33–34, 49, 99; tailings water, 58,
    61–68, 70–72, 74, 78, 87, 92, 95, 107, 112
Teniente: Mina El Teniente, 1, 31;
    Semanario El Teniente, 24–25;
    Teniente's environmental area, 60, 68;
    Teniente's waste management unit, 39,
    41, 44, 48, 54
territorialization, 40
toxicity, 13–14; toxicants, 3, 13, 99
toxicology, 13; politics of, 99
toxic torts, 98–99

ultimate sink, 11, 54
Uriarte, Ana Lya, 1

Valdebenito, 73, 107–8
Varas, Fabiola (pseud.), 36, 89–90
Venegas, Tomás (pseud.), 56, 59, 65

water: Water Code, 72, 77; water com-
    modification, 73; water rights, 72–74,
    77, 81

Zinc Corporation, 28–29

Founded in 1893,
UNIVERSITY OF CALIFORNIA PRESS
publishes bold, progressive books and journals
on topics in the arts, humanities, social sciences,
and natural sciences—with a focus on social
justice issues—that inspire thought and action
among readers worldwide.

The UC PRESS FOUNDATION
raises funds to uphold the press's vital role
as an independent, nonprofit publisher, and
receives philanthropic support from a wide
range of individuals and institutions—and from
committed readers like you. To learn more, visit
ucpress.edu/supportus.